OXFORD IB DIPLOMA PROGRAMME

MATHEMATICS HIGHER LEVEL:
CALCULUS

COURSE COMPANION

Josip Harcet
Lorraine Heinrichs
Palmira Mariz Seiler
Marlene Torres Skoumal

OXFORD
UNIVERSITY PRESS

Great Clarendon Street, Oxford OX2 6DP

Oxford University Press is a department of the University of Oxford.
It furthers the University's objective of excellence in research, scholarship,
and education by publishing worldwide in

Oxford   New York

Auckland Cape Town Dar es Salaam Hong Kong Karachi
Kuala Lumpur Madrid Melbourne Mexico City Nairobi
New Delhi Shanghai Taipei Toronto

With offices in

Argentina Austria Brazil Chile Czech Republic France Greece
Guatemala Hungary Italy Japan South Korea Poland Portugal
Singapore Switzerland Thailand Turkey Ukraine Vietnam

Oxford is a registered trade mark of Oxford University Press
in the UK and in certain other countries

© Oxford University Press 2014

The moral rights of the author have been asserted

Database right Oxford University Press (maker)

First published 2014

All rights reserved. No part of this publication may be reproduced,
stored in a retrieval system, or transmitted, in any form or by any means,
without the prior permission in writing of Oxford University Press,
or as expressly permitted by law, or under terms agreed with the appropriate
reprographics rights organization. Enquiries concerning reproduction
outside the scope of the above should be sent to the Rights Department,
Oxford University Press, at the address above

You must not circulate this book in any other binding or cover
and you must impose this same condition on any acquirer

British Library Cataloguing in Publication Data

Data available

ISBN 978-0-19-830484-5

10 9 8 7 6 5 4 3

Printed in Great Britain by Ashford Colour Press Ltd., Gosport

Paper used in the production of this book is a natural, recyclable product made from wood grown in sustainable forests.
The manufacturing process conforms to the environmental regulations of the country of origin.

Acknowledgements

The publisher would like to thank the following for permission to reproduce photographs:

p2: jupeart/Shutterstock; **p2:** mary416/Shutterstock; **p3:** freesoulproduction/Shutterstock; **p17:** Science Photo Library;
p21: Science Photo Library; **p39:** Scott Camazine/PRI/Getty.

Course Companion definition

The IB Diploma Programme Course Companions are resource materials designed to support students throughout their two-year Diploma Programme course of study in a particular subject. They will help students gain an understanding of what is expected from the study of an IB Diploma Programme subject while presenting content in a way that illustrates the purpose and aims of the IB. They reflect the philosophy and approach of the IB and encourage a deep understanding of each subject by making connections to wider issues and providing opportunities for critical thinking.

The books mirror the IB philosophy of viewing the curriculum in terms of a whole-course approach; the use of a wide range of resources, international mindedness, the IB learner profile and the IB Diploma Programme core requirements, theory of knowledge, the extended essay, and creativity, action, service (CAS).

Each book can be used in conjunction with other materials and indeed, students of the IB are required and encouraged to draw conclusions from a variety of resources. Suggestions for additional and further reading are given in each book and suggestions for how to extend research are provided.

In addition, the Course Companions provide advice and guidance on the specific course assessment requirements and on academic honesty protocol. They are distinctive and authoritative without being prescriptive.

IB mission statement

The International Baccalaureate aims to develop inquiring, knowledgable and caring young people who help to create a better and more peaceful world through intercultural understanding and respect.

To this end the IB works with schools, governments and international organizations to develop challenging programmes of international education and rigorous assessment.

These programmes encourage students across the world to become active, compassionate, and lifelong learners who understand that other people, with their differences, can also be right.

The IB learner Profile

The aim of all IB programmes is to develop internationally minded people who, recognizing their common humanity and shared guardianship of the planet, help to create a better and more peaceful world. IB learners strive to be:

Inquirers They develop their natural curiosity. They acquire the skills necessary to conduct inquiry and research and show independence in learning. They actively enjoy learning and this love of learning will be sustained throughout their lives.

Knowledgable They explore concepts, ideas, and issues that have local and global significance. In so doing, they acquire in-depth knowledge and develop understanding across a broad and balanced range of disciplines.

Thinkers They exercise initiative in applying thinking skills critically and creatively to recognize and approach complex problems, and make reasoned, ethical decisions.

Communicators They understand and express ideas and information confidently and creatively in more than one language and in a variety of modes of communication. They work effectively and willingly in collaboration with others.

Principled They act with integrity and honesty, with a strong sense of fairness, justice, and respect for the dignity of the individual, groups, and communities. They take responsibility for their own actions and the consequences that accompany them.

Open-minded They understand and appreciate their own cultures and personal histories, and are open to the perspectives, values, and traditions of other individuals and communities. They are accustomed to seeking and evaluating a range of points of view, and are willing to grow from the experience.

Caring They show empathy, compassion, and respect towards the needs and feelings of others. They have a personal commitment to service, and act to make a positive difference to the lives of others and to the environment.

Risk-takers They approach unfamiliar situations and uncertainty with courage and forethought, and have the independence of spirit to explore new roles, ideas, and strategies. They are brave and articulate in defending their beliefs.

Balanced They understand the importance of intellectual, physical, and emotional balance to achieve personal well-being for themselves and others.

Reflective They give thoughtful consideration to their own learning and experience. They are able to assess and understand their strengths and limitations in order to support their learning and personal development.

A note on academic honesty

It is of vital importance to acknowledge and appropriately credit the owners of information when that information is used in your work. After all, owners of ideas (intellectual property) have property rights. To have an authentic piece of work, it must be based on your individual and original ideas with the work of others fully acknowledged. Therefore, all assignments, written or oral, completed for assessment must use your own language and expression. Where sources are used or referred to, whether in the form of direct quotation or paraphrase, such sources must be appropriately acknowledged.

How do I acknowledge the work of others?

The way that you acknowledge that you have used the ideas of other people is through the use of footnotes and bibliographies.

Footnotes (placed at the bottom of a page) or endnotes (placed at the end of a document) are to be provided when you quote or paraphrase from another document, or closely summarize the information provided in another document. You do not need to provide a footnote for information that is part of a 'body of knowledge'. That is, definitions do not need to be footnoted as they are part of the assumed knowledge.

Bibliographies should include a formal list of the resources that you used in your work. 'Formal' means that you should use one of the several accepted forms of presentation. This usually involves separating the resources that you use into different categories (e.g. books, magazines, newspaper articles, Internet-based resources, CDs and works of art) and providing full information as to how a reader or viewer of your work can find the same information. A bibliography is compulsory in the extended essay.

What constitutes malpractice?

Malpractice is behaviour that results in, or may result in, you or any student gaining an unfair advantage in one or more assessment component. Malpractice includes plagiarism and collusion.

Plagiarism is defined as the representation of the ideas or work of another person as your own. The following are some of the ways to avoid plagiarism:

- Words and ideas of another person used to support one's arguments must be acknowledged.
- Passages that are quoted verbatim must be enclosed within quotation marks and acknowledged.
- CD-ROMs, email messages, web sites on the Internet, and any other electronic media must be treated in the same way as books and journals.
- The sources of all photographs, maps, illustrations, computer programs, data, graphs, audio-visual, and similar material must be acknowledged if they are not your own work.
- Words of art, whether music, film, dance, theatre arts, or visual arts, and where the creative use of a part of a work takes place, must be acknowledged.

Collusion is defined as supporting malpractice by another student. This includes:

- allowing your work to be copied or submitted for assessment by another student
- duplicating work for different assessment components and/or diploma requirements.

Other forms of malpractice include any action that gives you an unfair advantage or affects the results of another student. Examples include, taking unauthorized material into an examination room, misconduct during an examination, and falsifying a CAS record.

About the book

The new syllabus for Mathematics Higher Level Option: Calculus is thoroughly covered in this book. Each chapter is divided into lesson-size sections with the following features:

The Course Companion will guide you through the latest curriculum with full coverage of all topics and the new internal assessment. The emphasis is placed on the development and improved understanding of mathematical concepts and their real life application as well as proficiency in problem solving and critical thinking. The Course Companion denotes questions that would be suitable for examination practice and those where a GDC may be used.

Questions are designed to increase in difficulty, strengthen analytical skills and build confidence through understanding.

Where appropriate the solutions to examples are given in the style of a graphics display calculator.

Mathematics education is a growing, ever changing entity. The contextual, technology integrated approach enables students to become adaptable, lifelong learners.

Note: US spelling has been used, with IB style for mathematical terms.

About the authors

Lorraine Heinrichs has been teaching mathematics for 30 years and IB mathematics for the past 16 years at Bonn International School. She has been the IB DP coordinator since 2002. During this time she has also been senior moderator for HL Internal Assessment and workshop leader of the IB; she was also a member of the curriculum review team.

Palmira Mariz Seiler has been teaching mathematics for over 25 years. She joined the IB community in 2001 as a teacher at the Vienna International School and since then has also worked as Internal Assessment moderator in curriculum review working groups and as a workshop leader and deputy chief examiner for HL mathematics. Currently she teaches at Colegio Anglo Colombiano in Bogota, Colombia.

Marlene Torres-Skoumal has taught IB mathematics for over 30 years. During this time, she has enjoyed various roles with the IB, including deputy chief examiner for HL, senior moderator for Internal Assessment, calculator forum moderator, workshop leader, and a member of several curriculum review teams.

Josip Harcet has been involved with and teaching the IB programme since 1992. He has served as a curriculum review member, deputy chief examiner for Further Mathematics, assistant IA examiner and senior examiner for Mathematics HL as well as a workshop leader since 1998.

Contents

Chapter 1 Patterns to infinity   **2**
1.1 From limits of sequences to limits of functions   3
1.2 Squeeze theorem and the algebra of limits of convergent sequences   7
1.3 Divergent sequences: indeterminate forms and evaluation of limits   10
1.4 From limits of sequences to limits of functions   13

Chapter 2 Smoothness in mathematics   **22**
2.1 Continuity and differentiability on an interval   24
2.2 Theorems about continuous functions   28
2.3 Differentiable functions: Rolle's Theorem and Mean Value Theorem   33
2.4 Limits at a point, indeterminate forms, and L'Hopital's rule   42
2.5 What are *smooth graphs* of functions?   49
2.6 Limits of functions and limits of sequences   50

Chapter 3 Modeling dynamic phenomena   **54**
3.1 Classifications of differential equations and their solutions   56
3.2 Differential Equations with separated variables   61
3.3 Separable variables, differential equations and graphs of their solutions   63
3.4 Modeling of growth and decay phenomena   69
3.5 First order exact equations and integrating factors   73
3.6 Homogeneous differential equations and substitution methods   80
3.7 Euler Method for first order differential equations   85

Chapter 4 The finite in the infinite   **96**
4.1 Series and convergence   98
4.2 Introduction to convergence tests for series   104
4.3 Improper Integrals   110
4.4 Integral test for convergence   112
4.5 The p-series test   114
4.6 Comparison test for convergence   115
4.7 Limit comparison test for convergence   118
4.8 Ratio test for convergence   119
4.9 Absolute convergence of series   120
4.10 Conditional convergence of series   122

Chapter 5 Everything polynomic   **130**
5.1 Representing Functions by Power Series 1   132
5.2 Representing Power Series as Functions   135
5.3 Representing Functions by Power Series 2   138
5.4 Taylor Polynomials   143
5.5 Taylor and Maclaurin Series   146
5.6 Using Taylor Series to approximate functions   156
5.7 Useful applications of power series   161

Answers   **168**

Index   **185**

1 Patterns to infinity

CHAPTER OBJECTIVES:

9.1 Infinite sequences of real numbers and their convergence or divergence. Informal treatment of limit of sum, difference, product, quotient, squeeze theorem.

Before you start

You should know how to:

1. Simplify algebraic fractions.
   e.g. $\dfrac{1}{n} - \dfrac{1}{n+1} = \dfrac{n+1-n}{n(n+1)} = \dfrac{1}{n(n+1)}$

2. Solve inequalities that involve absolute value.
   e.g. Solve $|x - 3| < 2$
   $|x - 3| < 2 \Rightarrow -2 < x - 3 < 2 \Rightarrow 1 < x < 5$

3. Rationalise denominators of fractions.
   e.g. $\dfrac{3}{\sqrt{n} - \sqrt{n+2}} = \dfrac{3(\sqrt{n} + \sqrt{n+2})}{(\sqrt{n} - \sqrt{n+2})(\sqrt{n} + \sqrt{n+2})}$
   $= \dfrac{3(\sqrt{n} + \sqrt{n+2})}{n - (n+2)} = -\dfrac{3(\sqrt{n} + \sqrt{n+2})}{2}$

4. Recognise arithmetic and geometric sequences and use knowledge about their general terms to determine the expression of other sequences.
   e.g. Find the nth term of the sequence
   $\dfrac{1}{3}, -\dfrac{2}{5}, \dfrac{4}{7}, -\dfrac{8}{9}, \dfrac{16}{11}, \ldots$
   Each term can be seen as the quotient of terms of a geometric sequence (numerator) and the quotient of terms of an arithmetic sequence (denominator). The general term of this sequence is
   $u_n = \dfrac{(-2)^{n-1}}{2n+1}, n \in \mathbb{Z}^+$

Skills check:

1. Simplify:
   a $\dfrac{3}{4n+1} - \dfrac{3}{4n-1}$
   b $\dfrac{n-1}{n+2} - \dfrac{n}{n+3}$

2. Solve:
   a $|3x - 1| < 1$
   b $|2x - 3| \geq 4$

3. Rationalise the denominator:
   a $\dfrac{2n}{1 + \sqrt{n}}$
   b $\dfrac{n-1}{\sqrt{n} + \sqrt{n+1}}$

4. Find the nth term of
   a $\dfrac{1}{2}, -\dfrac{1}{4}, \dfrac{1}{8}, -\dfrac{1}{16}, \ldots$
   b $\dfrac{1}{2}, -\dfrac{1}{3}, \dfrac{1}{4}, -\dfrac{1}{5}, \ldots$
   c $\dfrac{1}{2}, \dfrac{3}{4}, \dfrac{5}{6}, \dfrac{7}{8}, \ldots$
   d $\dfrac{\sqrt{3}}{2}, -\dfrac{\sqrt{5}}{4}, \dfrac{\sqrt{7}}{8}, -\dfrac{3}{16}, \ldots$

1.1 From limits of sequences to limits of functions

Infinity is a concept that has challenged mathematicians and scientists for centuries. Throughout this time the concept of infinity was sometimes denied and sometimes accepted by mathematicians, to the point that it became one of the main issues in the history of Mathematics. In the last 150 years, great advances were made: first with the axiomatization of set theory; and then with the work of philosopher Bertrand Russell and his collection of paradoxes. At the climax of all discussions was the work of Georg Cantor on the classification of infinities.

But do infinities really exist? After all, how many types of infinity are there? Does it make sense to compare them and operate with them?

In this chapter we will explore the concept of infinity, starting with an intuitive approach and looking at familiar number patterns: sequences. We will then formalise the idea of 'the pattern that goes on forever' and formally define the limit of a sequence. This may help you to better understand the theorems about sequences, although formal treatment of limits of sequences will not be examined. For this reason, all proofs of results have been omitted.

> Intuitively we think of infinity as something larger than any number. This sounds simple, but in fact infinity is anything but simple. In the late 1800s, Georg Cantor discovered that there are many different sizes of infinity, in fact there are infinitely many sizes! Cantor showed that the smallest infinity is the one you would get to if you could count forever (a 'countable' infinity), and that infinities which aren't countable, such as real numbers, are actually larger in size.

> Is there a 'medium' size of infinity, bigger than the natural numbers but smaller than the real numbers? The supposition that nothing is in between the two infinities is called the 'continuum hypothesis'. You may want to explore this topic further to learn about the powerful methods invented by Cantor and the resulting problems which still puzzle mathematicians today.

At the end of the chapter we will explore the connections between limits of sequences and limits of functions introduced in the core course. We will also establish criteria for the existence of the limit of a function at a point.

Consider the following numerical sequences:

a_n: 1, 2, 4, 8, 16, ..., 2^n, ...

b_n: 1, $\frac{1}{2}$, $\frac{1}{3}$, $\frac{1}{4}$, ..., $\frac{1}{n}$, ...

c_n: −1, 1, −1, 1, −1, 1, ..., $(-1)^n$, ...

What is happening to the terms of these sequences as n increases?
Do they approach any real number as $n \to +\infty$?

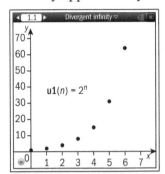
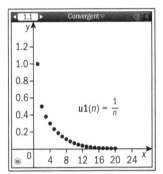
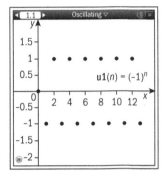

If we graph the sequences $\{a_n\}$, $\{b_n\}$, and $\{c_n\}$, we can observe their behaviour as n increases, and notice that:

- $\lim_{n \to \infty} a_n = +\infty$ which means that $\{a_n\}$ diverges;
- $\lim_{n \to \infty} b_n = 0$ which means that $\{b_n\}$ converges to 0;
- $\lim_{n \to \infty} c_n$ does not exist (it oscillates from 1 to −1) which means that $\{c_n\}$ diverges.

The following investigation will help you to better understand what 'convergent' means.

Investigation 1

1 Use technology to graph the sequence defined by $u_n = \dfrac{n+1}{2n+1}$.

2 Hence explain why $\lim_{n \to \infty} u_n = \dfrac{1}{2}$.

3 Find the minimum value of m such that $n \geq m \Rightarrow \left| u_n - \dfrac{1}{2} \right| < 0.1$ (i.e. find the smallest integer n for which the difference between the value u_n and $\dfrac{1}{2}$ is less than 0.1).

4 Consider the positive small quantities $\varepsilon = 0.01$, 0.001, and 0.0001. In each case find the minimum value of m such that $n \geq m \Rightarrow \left| u_n - \dfrac{1}{2} \right| < \varepsilon$.

5 Decide whether or not it is possible to find the order m such that $n \geq m \Rightarrow |u_n - 0.4| < 0.1$. Give reasons for your answer.

Consider now the sequence defined by $v_n = \left(-\dfrac{1}{3} \right)^n$.

6 Explain why $\lim_{n \to \infty} v_n = 0$.

7 Consider the positive small quantities $\varepsilon = 0.01$, 0.001, and 0.0001. In each case find the minimum value of m such that $n \geq m \Rightarrow |v_n| < \varepsilon$.

8 Explain the meaning of $\lim_{n \to \infty} u_n = L$ in terms of the value of $|u_n - L|$.

> You may want to use sequences defined by expressions involving arithmetic and geometric sequences studied as part of the core course.

9 Explore further cases of your choice.

Definition: Convergent sequences

$\{u_n\}$ is a convergent sequence with $\lim_{n\to\infty} u_n = L$ if and only if for any $\varepsilon > 0$ there exists a least order $m \in \mathbb{Z}^+$ such that, for all $n \geq m \Rightarrow |u_n - L| < \varepsilon$.

> The Greek letter epsilon, ε, is used among mathematicians all over the world to represent a small positive quantity.

This definition gives an algebraic criterion to test whether or not a given number L is the limit of a sequence. However, to apply this test, you must first decide about the value of L.

Example 1

Show that the sequence defined by $u_n = \dfrac{3n-1}{n+1}$ is convergent.

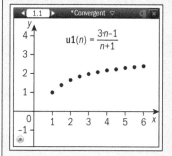

Graph the sequence and observe its behavior as n increases.

$|u_n - 3| = \left|\dfrac{3n-1}{n+1} - 3\right| = \left|\dfrac{3n-1-3n-3}{n+1}\right| = \dfrac{4}{n+1}$

Find a simplified expression for $|u_n - 3|$

$|u_n - 3| < \varepsilon \Rightarrow \dfrac{4}{n+1} < \varepsilon \Rightarrow n+1 > \dfrac{4}{\varepsilon} \Rightarrow n > \dfrac{4}{\varepsilon} - 1$

The value of m is the least positive integer greater than $\dfrac{4}{\varepsilon} - 1$

So, it is possible to find m such that
$n \geq m \Rightarrow |u_n - L| < \varepsilon$.
$\therefore \lim_{n\to+\infty} u_n = 3$

Use the definition to show that $\lim_{n\to\infty} u_n = 3$.

Note that this definitions tells you that from the order m onwards, all the terms of the sequence lie within the interval $]L - \varepsilon, L + \varepsilon[$. This means that the sequence can have exactly one limit, L.

Useful theorems about subsequences of convergent and divergent sequences:

- If $\{b_n\} \subseteq \{a_n\}$ is a subsequence of a convergent sequence $\{a_n\}$, then $\{b_n\}$ is also a convergent sequence and $\lim_{n\to\infty} b_n = \lim_{n\to\infty} a_n$
- If $\{b_n\} \subseteq \{a_n\}$ and $\{c_n\} \subseteq \{a_n\}$ are subsequences of a sequence $\{a_n\}$ and $\lim_{n\to\infty} b_n \neq \lim_{n\to\infty} c_n$ then $\{a_n\}$ is not convergent (i.e. $\{a_n\}$ is a divergent sequence).

> It is usual to use set notation when describing sequences and their subsequences. For example $\{b_n\} \subseteq \{a_n\}$ means that the sequence $\{b_n\}$ can be obtained from the sequence $\{a_n\}$ by removing at least one of the terms of $\{a_n\}$. In this way, $\{b_n\}$ can be seen as a part of $\{a_n\}$.

The following examples show you how to use subsequences of a given sequence to show that the sequence diverges.

Example 2

Show that the sequence defined by $a_n = (-1)^n \cdot 2$ does not converge.	
$a_n: -2, 2, -2, 2, \ldots$	Calculate a few terms of the sequence and observe the pattern.
$b_n = a_{2n} = 2 \to 2$ and $c_n = a_{2n-1} = -2 \to -2$	$\{b_n\}$ and $\{c_n\}$ are the subsequences of even order terms, and odd order terms, respectively.
$\therefore \lim_{n\to\infty} b_n \neq \lim_{n\to\infty} c_n$ then $\{a_n\}$ is not convergent.	If the sequence $\{a_n\}$ was convergent all its subsequences would have the same limit.

Example 3

Show that the sequence defined by $u_n: \begin{cases} u_1 = 3 \\ u_{n+1} = -u_n \end{cases}$, $n \in \mathbb{Z}^+$ is divergent.	
Let $\{v_n\}$ be the subsequence of $\{u_n\}$ of the terms of even order, i.e. $v_n = u_{2n} = -3$ (m is any positive integer), and let $\{w_n\}$ be the subsequence of $\{u_n\}$ of the terms of odd order, i.e. $w_n = u_{2m-1} = 3$.	$\{u_n\}$ and $\{v_n\}$ are the subsequences of even order terms, and odd order terms, respectively. $3, -3, 3, -3, \ldots$
Since $\lim_{n\to\infty} v_n = -3 \neq 3 = \lim_{n\to\infty} w_n$ the sequence $\{u_n\}$ cannot converge.	If the sequence $\{u_n\}$ was convergent all its subsequences would have the same limit.

We can also use subsequences to determine the limit of a convergent sequence defined recursively as we will see later in this chapter.

Exercise 1A

1 Consider the sequence defined by $u_n = \dfrac{n+3}{2n+1}$. Find the least value of $m \in \mathbb{Z}^+$ such that $n \geq m \Rightarrow \left| u_n - \dfrac{1}{2} \right| < 0.001$.

2 Consider the sequence defined by $v_n = \dfrac{n+1}{3n-1}$.
   a Graph the sequence and, if possible, state its limit.
   b Find the least value of $m \in \mathbb{Z}^+$ such that $n > m \Rightarrow \left| v_n - \dfrac{1}{3} \right| < 0.001$

3 Consider the sequence defined by $u_n = \dfrac{4^n - 3}{4^n}$.
   a Graph the sequence and, if possible, state its limit.
   b Find the least value of $m \in \mathbb{Z}^+$ such that $n > m \Rightarrow |u_n - 1| < 0.0005$

4 State whether or not the following sequences are convergent, giving reasons for your answers:

a $a_n = (-1)^n \dfrac{3n}{4n+1}$

b $b_n = \left(-\dfrac{1}{n}\right)^n$

c $c_n = \left(\dfrac{1}{n}\right)^{(-1)^n}$

d $d_n = \sqrt{n+1} - \sqrt{n}$

e $e_n = n^3 - n^2$

f $f_n = \sin\left(\dfrac{n\pi}{2}\right)$

g $g_n = \begin{cases} \cos(n\pi), & n \text{ even} \\ \dfrac{1}{n}, & n \text{ odd} \end{cases}$

h $h_n = \dfrac{n^3 + n + 1}{2n^3 + 3}$

> When all the terms of a sequence $\{u_n\}$ take values between two real numbers M and N, i.e. $M \leq u_n \leq N$ for all $n \in \mathbb{Z}^+$, we say that the sequence is bounded. M and N are then called lower and upper bounds, respectively.

5 Use the result 'if $\{u_n\}$ is bounded and $\lim_{n\to\infty} v_n = 0$ then $\lim_{n\to\infty} u_n \cdot v_n = 0$', to show that $\lim_{n\to\infty} \dfrac{n\cos^2(na)}{n^2 + 3} = 0$, $a \in \mathbb{R}$.

1.2 Squeeze theorem and the algebra of limits of convergent sequences

The following result is a very useful theorem to calculate limits of sequences.

Theorem 1 (Squeeze Theorem): Consider three sequences such that:

- There exists some $p \in \mathbb{Z}^+$ such that $u_n \leq v_n \leq w_n$ for all $n \geq p \in \mathbb{Z}^+$
- $\{u_n\}$ and $\{w_n\}$ converge and $\lim_{n\to\infty} u_n = \lim_{n\to\infty} w_n = L$

Then $\{v_n\}$ converges and $\lim_{n\to\infty} v_n = L$

> Squeeze theorem is also known as *Sandwich Theorem*.

Example 4

Prove that the sequence defined by $v_n = \dfrac{\sin(2n+1)}{n}$ is convergent.	
$\dfrac{-1}{n} \leq \dfrac{\sin(2n+1)}{n} \leq \dfrac{1}{n}$, $n \in \mathbb{Z}^+$ and	$-1 \leq \sin(2n+1) \leq 1$ for all $n \in \mathbb{Z}^+$
$\lim_{n\to\infty}\left(-\dfrac{1}{n}\right) = \lim_{n\to\infty}\left(\dfrac{1}{n}\right) = 0 \Rightarrow \lim_{n\to\infty} \underbrace{\left(\dfrac{\sin(2n+1)}{n}\right)}_{v_n} = 0$	Apply the Squeeze Theorem.
$\therefore \{v_n\}$ converges.	

The next result establishes formally the algebra of limits of sequences.

Theorem 2: Let $\{u_n\}$ and $\{v_n\}$ be convergent sequences and $\lim\limits_{n\to\infty} u_n = L_1$ and $\lim\limits_{n\to\infty} v_n = L_2$. Then:

i $\lim\limits_{n\to\infty}(u_n + v_n) = L_1 + L_2$ **ii** $\lim\limits_{n\to\infty}(u_n - v_n) = L_1 - L_2$

iii $\lim\limits_{n\to\infty}(u_n \cdot v_n) = L_1 L_2$ **iv** $\lim\limits_{n\to\infty}\left(\dfrac{u_n}{v_n}\right) = \dfrac{L_1}{L_2}$ when $L_2 \neq 0$.

Example 5

Prove that the sequence defined by $u_n = \left(3 + \dfrac{2}{n}\right)\left(2 + \dfrac{1}{n+1}\right)^2$ converges to 12.

$\lim\limits_{n\to\infty} u_n = \lim\limits_{n\to\infty}\left(3 + \dfrac{2}{n}\right) \cdot \lim\limits_{n\to\infty}\left(2 + \dfrac{1}{n+1}\right)^2$

$= \left(\lim\limits_{n\to\infty} 3 + \lim\limits_{n\to\infty}\dfrac{2}{n}\right) \cdot \left(\lim\limits_{n\to\infty} 2 + \lim\limits_{n\to\infty}\dfrac{1}{n+1}\right) \cdot \left(\lim\limits_{n\to\infty} 2 + \lim\limits_{n\to\infty}\dfrac{1}{n+1}\right)$

$= (3 + 0)\cdot(2 + 0)\cdot(2 + 0) = 12$

As a consequence of Theorem 2 (iii), if k is a positive integer, $\left(\lim\limits_{n\to\infty}(u_n)^k\right) = \left(\lim\limits_{n\to\infty} u_n\right)^k$

Apply theorem 2 to reduce the calculation to simpler cases:

$\lim\limits_{n\to\infty}\dfrac{2}{n} = \lim\limits_{n\to\infty}\dfrac{1}{n+1} = 0$ *and*

$\lim\limits_{n\to\infty} 3 = 3$ *and* $\lim\limits_{n\to\infty} 2 = 2$

The following example shows you how to use subsequences and properties of limits of sequences to find limits of sequences defined recursively:

Example 6

Given the convergent sequence defined by
$u_n : \begin{cases} u_1 = 1 \\ u_{n+1} = 1 + \dfrac{1}{u_n}, \quad n \in \mathbb{Z}^+ \end{cases}$
find its limit.

Let L be the limit of $\{u_n\}$. As the sequence $\{u_n\}$ is convergent, its subsequence $\{u_{n+1}\}$ has the same limit L.

$\lim\limits_{n\to\infty} u_{n+1} = \lim\limits_{n\to\infty}\left(1 + \dfrac{1}{u_n}\right) \Rightarrow L = 1 + \dfrac{1}{L}$

$\Rightarrow L^2 - L - 1 = 0 \Rightarrow L = \dfrac{1 \pm \sqrt{5}}{2}$

As all the terms of $\{u_n\}$ are positive,

$L = \dfrac{1 + \sqrt{5}}{2}$.

Calculate a few terms of the sequence $\{u_n\}$:

1, 2, $\dfrac{3}{2}$, $\dfrac{5}{3}$, $\dfrac{8}{3}$, ... and notice that all the terms are positive; you may also notice that each term is equal to the quotient of two consecutive Fibonacci numbers.

Apply Theorem 2 to obtain a quadratic equation in L.

Solve the equation and find all possible values of L.

Use the fact that all terms of the sequence are positive to eliminate one of the possible values of L and conclude that L is the golden ratio.

Exercise 1B

1 Show that following sequences converge to zero:

a $a_n = 2 \cdot (-1)^{n-1} \cdot \left(\dfrac{3}{5}\right)^n$

b $b_n = \dfrac{\cos^2(3n)}{3n+1}$

2 Consider the sequence defined by $u_n = \dfrac{3n + \sin(2n)}{4n - 3},\ n \in \mathbb{Z}^+$.

a Show that $\dfrac{3n-1}{4n-3} \leq \dfrac{3n+\sin(2n)}{4n-3} \leq \dfrac{3n+1}{4n-3}$ for all $n \in \mathbb{Z}^+$.

b Find $\lim\limits_{n \to \infty} \dfrac{3n-1}{4n-3}$ and $\lim\limits_{n \to \infty} \dfrac{3n+1}{4n-3}$

c Use the squeeze theorem to find $\lim\limits_{n \to \infty} u_n$.

3 Consider the sequence defined by $u_n = \left(\dfrac{2n}{3n+1}\right)^n,\ n \in \mathbb{Z}^+$.

a Show that $\dfrac{1}{2} \leq \dfrac{2n}{3n+1} < \dfrac{2}{3}$ for all $n \in \mathbb{Z}^+$.

b Hence use the squeeze theorem to find $\lim\limits_{n \to \infty} u_n$.

4 Consider the sequence defined by the following recurrence formula:

$$\begin{cases} a_1 = 1 \\ a_{n+1} = \dfrac{a_n + 2}{3}, & n \in \mathbb{Z}^+ \end{cases}$$

Use the result $\lim\limits_{n \to \infty} a_{n+1} = \lim\limits_{n \to \infty} a_n$ to find the limit of the sequence. State any assumption you need to make to calculate this limit.

1.3 Divergent sequences: indeterminate forms and evaluation of limits

In this section you are going to learn more about limits of divergent sequences that are not bounded and how to operate with these limits. You will be introduced to the algebra of infinity and learn more about the indeterminate forms studied in the core course.

Investigation 2

1. Use technology to graph the sequence defined by $u_n = 3^{n-1}$.
2. Hence explain why $\lim_{n \to \infty} u_n = +\infty$.
3. Find the minimum value of m such that $n \geq m \Rightarrow u_n > 100$, i.e. find the least order for which the all the terms of the sequence are greater than or equal to 100.
4. Consider large positive quantities $L = 1000, 10\,000$, and $1\,000\,000$. In each case find the least order m such that $n \geq m \Rightarrow u_n \geq L$.
5. Consider now the sequence defined by $v_n = -2^n$. Consider large positive quantities $L = 1000, 10\,000$, and $1\,000\,000$. In each case find the least order m such that $n \geq m \Rightarrow v_n \leq -L$.
6. Explain why $\lim_{n \to \infty} v_n = -\infty$.
7. Let $w_n = (-4)^n$. Consider large positive quantities $L = 1000, 10\,000$, and $1\,000\,000$. In each case find the least order m such that $n \geq m \Rightarrow |w_n| \geq L$.
8. Explain the meaning of $\lim_{n \to \infty} |w_n| = +\infty$ in terms of the value of $|w_n|$.
9. Explore further cases of your choice.

> You may want to use sequences defined by expressions involving arithmetic and geometric sequences studied as part of the core course.

Definitions: Given a sequence $\{u_n\}$, if for all given $L > 0$, there exists an order $m \in \mathbb{N}^+$ such that:

i $n \geq m \Rightarrow u_n > L$ then this means that $\lim_{n \to \infty} u_n = +\infty$

ii $n \geq m \Rightarrow u_n < -L$ then this means that $\lim_{n \to \infty} u_n = -\infty$

iii $n \geq m \Rightarrow |u_n| > L$ then this means that $\lim_{n \to \infty} |u_n| = +\infty$

Note that if $\lim_{n \to \infty} u_n = +\infty$ or $\lim_{n \to \infty} u_n = -\infty$ then $\lim_{n \to \infty} |u_n| = +\infty$ but the converse may not be true. For example, for $u_n = (-3)^n$, $\lim_{n \to \infty} |u_n| = +\infty$ but neither $\lim_{n \to \infty} u_n = +\infty$ nor $\lim_{n \to \infty} u_n = -\infty$.

Example 7

Prove that

a $\lim\limits_{n\to\infty}(2n-1) = +\infty$ **b** $\lim\limits_{n\to\infty}(-3^n+5) = -\infty$ **c** $\lim\limits_{n\to\infty}|(-2)^n| = +\infty$

Given $L > 0$,

a As $2n - 1 > L \Rightarrow n > \dfrac{L+1}{2}$, for $p = \text{int}\left(\dfrac{L+1}{2}\right) + 1$, it is true that $n \geq p \Rightarrow u_n > L$. Therefore $\lim\limits_{n\to\infty} u_n = +\infty$.

b As $-3^n + 5 < -L \Rightarrow 3^n > L + 5 \Rightarrow n > \log_3(L+5)$, for $p = \text{int}(\log_3(L+5)) + 1$, it is true that $n \geq p \Rightarrow u_n < -L$. Therefore $\lim\limits_{n\to\infty} u_n = -\infty$.

c As $|(-2)^n| > L \Rightarrow 2^n > L \Rightarrow n > \log_2(L)$, for $p = \text{int}(\log_2(L)) + 1$, it is true that $n \geq p \Rightarrow |u_n| > L$. Therefore $\lim\limits_{n\to\infty}|u_n| = +\infty$.

> $p = \text{int}(x)$ means the largest integer p, smaller than or equal to x
>
> Use definitions of limits of unbounded sequences.

The following important and useful theorem relates sequences that have limit infinity with convergent sequences that have limit zero.

Theorem 3: Let $\{u_n\}$ be a sequence.

i $\lim\limits_{n\to\infty}|u_n| = +\infty \Rightarrow \lim\limits_{n\to\infty}\dfrac{1}{u_n} = 0$

ii $\lim\limits_{n\to\infty} u_n = 0 \Rightarrow \lim\limits_{n\to\infty}\left|\dfrac{1}{u_n}\right| = +\infty$

This theorem formalizes results that you have been using intuitively: the reciprocal of something infinitely large is something infinitely small, and the reciprocal of something infinitely small is something infinitely large. This is important when it comes to setting rules for the algebra of infinity. For example, using Theorem 3, it is easy to prove that if $\lim\limits_{x\to\infty} u_n = a \neq 0$ and $\lim\limits_{x\to\infty} v_n = 0$, then $\lim\limits_{n\to\infty}\left(\dfrac{u_n}{v_n}\right) = \dfrac{a}{0} = +\infty$.

The notation $\dfrac{a}{0}$ needs to be interpreted in the context of limits and seen as a simplification of mathematical language. Using this simplified language, the algebra of limits involving infinity can be summarized by the following table:

$(\pm\infty) + (\pm\infty) = \pm\infty$	$(\pm\infty) \times (\pm\infty) = +\infty$	$a + (\pm\infty) = \pm\infty, a \in \mathbb{R}$
$(\pm\infty) - (\pm\infty) = $ indeterminate	$(\pm\infty) \times (\mp\infty) = \mp\infty$	$a - (\pm\infty) = \mp\infty, a \in \mathbb{R}$
$a \times (\pm\infty) = \pm\infty, a > 0$	$a \times (\pm\infty) = \mp\infty, a < 0$	$\infty^n = \infty, n \in \mathbb{Z}^+$
$\dfrac{a}{\infty} = 0, a \in \mathbb{R}$	$\dfrac{0}{0} = $ indeterminate	$\dfrac{\infty}{\infty} = $ indeterminate
$\dfrac{\infty}{a} = \infty, a \in \mathbb{R}, a \neq 0$	$\sqrt[n]{+\infty} = +\infty, n$ even $\\ \sqrt[n]{\pm\infty} = \pm\infty, n$ odd	$0 \times \infty = $ indeterminate

While most results shown on the table may appear trivial, it is very important that you recognize the indeterminate forms, i.e. situations for which there is no general rule and that you need to analyse case by case. In such instances, you will usually simplify or transform the expressions into an equivalent form. Examples are shown below.

Example 8

Show that the rules of the algebra of limits cannot be applied to evaluate the following limits. Then, manipulate the expressions and find their values.

a $\lim_{n \to \infty} \dfrac{4n^2 + 1}{3n^2 - 1}$ **b** $\lim_{n \to \infty} \left(\sqrt{n+1} - \sqrt{n}\right)$

'ind' stands for 'indeterminate form'

---

a $\lim_{n \to \infty} \dfrac{4n^2 + 1}{3n^2 - 1} = \dfrac{+\infty + 1}{+\infty - 1} = \dfrac{+\infty}{+\infty}$ (ind)

Apply algebra of limits rules and identify the indeterminate form.

$\lim_{n \to \infty} \dfrac{4n^2 + 1}{3n^2 - 1} = \lim_{n \to \infty} \dfrac{n^2\left(4 + \dfrac{1}{n^2}\right)}{n^2\left(3 - \dfrac{1}{n^2}\right)} = \dfrac{4 + 0}{3 - 0} = \dfrac{4}{3}$

Transform the expression by dividing both terms of the fraction by the highest power of n, and re-apply the algebra of limits rules.

b $\lim_{n \to \infty} \left(\sqrt{n+1} - \sqrt{n}\right) = \sqrt{+\infty} - \sqrt{+\infty} = (+\infty) - (+\infty)$ (ind)

$\lim_{n \to \infty} \left(\sqrt{n+1} - \sqrt{n}\right) = \lim_{n \to \infty} \dfrac{\left(\sqrt{n+1} - \sqrt{n}\right)\left(\sqrt{n+1} + \sqrt{n}\right)}{\left(\sqrt{n+1} + \sqrt{n}\right)}$

Use $(x - y)(x + y) = x^2 - y^2$

$= \lim_{x \to \infty} \dfrac{(n+1) - n}{\sqrt{n+1} + \sqrt{n}} = \lim_{x \to \infty} \dfrac{1}{\sqrt{n+1} + \sqrt{n}} = \dfrac{1}{+\infty} = 0$

Investigation 3

Use technology to calculate the value of the following expressions for large values of n. Then write your conjectures for the values of their limits, test them further and write your conclusions.

a $a_n = \dfrac{n}{2^n}$ **b** $b_n = \dfrac{n^k}{b^n}$ where $k \in \mathbb{Z}^+$, $b > 1$ **c** $c_n = \sqrt[n]{a}$, $a > 0$ **d** $d_n = \sqrt[n]{n}$ **e** $e_n = \sqrt[n]{n!}$

Exercise 1C

1 Evaluate the following limits:

a $\lim\limits_{n \to \infty} \dfrac{5n^3 - n}{2n^3 + 6n^2 - 1}$ **b** $\lim\limits_{n \to \infty} \dfrac{n^2 + 5n}{n^3 + 2n + 3}$ **c** $\lim\limits_{n \to \infty} \dfrac{n^2 + 6n - 1}{n + 1}$ **d** $\lim\limits_{n \to \infty} \left(n^3 - n\right)$

e $\lim\limits_{n \to \infty} \left(\sqrt{n + 5} - \sqrt{n + 2}\right)$ **f** $\lim\limits_{n \to \infty} \left(\sqrt{\dfrac{n + 5}{n + 2}}\right)$ **g** $\lim\limits_{n \to \infty} \dfrac{3^n + 2^n}{4^n + 5 \cdot 3^n}$ **h** $\lim\limits_{n \to \infty} \dfrac{(-3)^{n+1} + 7^n}{4^{n-1} + e^n}$

i $\lim\limits_{n \to \infty} \dfrac{\sqrt{n} - \sqrt[3]{n}}{\sqrt{n}}$ **j** $\lim\limits_{n \to \infty} \left(\dfrac{n^3}{n+1} - \dfrac{n^3 + 1}{n}\right)$ **k** $\lim\limits_{n \to \infty} \dfrac{\sum\limits_{k=0}^{n}(2k - 1)}{2n^2 + 1}$ **l** $\lim\limits_{n \to \infty} \dfrac{\sum\limits_{k=0}^{n}\left(2^{k-1}\right)}{3^n}$

EXTENSION QUESTION

2 Given that $\lim\limits_{n \to \infty} \sqrt[n]{n} = 1$,

a Show that $\lim\limits_{n \to \infty} \dfrac{\ln n}{n} = 0$. Hence find the value of $\lim\limits_{n \to +\infty} \dfrac{\ln(n + k)}{n}$, where $k > 0$.

b Find the value of $\lim\limits_{n \to \infty} \sqrt[n]{n \cdot \left(\dfrac{5}{n}\right)^n}$

1.4 From limits of sequences to limits of functions

During the 19th century, several mathematicians worked on concepts of calculus in an attempt to make this area of mathematics more rigorous. As the study of limits of functions forced mathematicians to deal with the concept of infinity, and with intuition often leading to incorrect results, it was very important to develop clear definitions. Cauchy was among this group of mathematicians, and he was the first to define the limit of a function precisely, using terms very similar to the ones we use today. He interpreted $\lim\limits_{x \to a} f(x) = b$ as a relation between variables: when the difference $\Delta x = x - a$ becomes infinitely small, the difference $\Delta y = f(x) - b$ also becomes infinitely small.

Over the course of the following 50 years, other definitions of limits of functions were proposed by various mathematicians. These were explored in order to deal with the pitfalls of Cauchy's definition which were exposed. Among them, there is a definition of limit of function known as Heine's definition, which is based on the study of limits of numerical sequences:

Definition: Let I be an open interval of real numbers. Let $f : I \to \mathbb{R}$ and $a \in I$, then $\lim_{x \to a} f(x)$ exists and $\lim_{x \to a} f(x) = b$ if and only if for any sequence $\{a_n\}$ such that $a_n \in I$ for all $n \in \mathbb{Z}^+$ and $\lim_{n \to \infty} a_n = a$, $\lim_{n \to \infty} (f(a_n)) = b$.

> This definition will be very useful for studying results in Chapter 2

This definition is very useful when we want to show that the limit of a function at a point $x = a$ does not exist. In this case, you just need to find two sequences $\{u_n\}$ and $\{v_n\}$ such that $\lim_{n \to +\infty} u_n = \lim_{n \to +\infty} v_n = a$ but $\lim_{n \to +\infty} (f(u_n)) \neq \lim_{n \to +\infty} (f(v_n))$.

Example 9

Show that the function f defined by $f(x) = \begin{cases} 3x - 1, & x < 2 \\ x - 2, & x \geq 2 \end{cases}$ has no limit at $x = 2$.

Let $u_n = 2 - \dfrac{1}{n}$ and $v_n = 2 + \dfrac{1}{n}$.

$\lim_{n \to +\infty} u_n = \lim_{n \to +\infty} v_n = 2$

$\lim_{n \to +\infty} (f(u_n)) = \lim_{n \to +\infty} \left(3\left(2 - \dfrac{1}{n}\right) - 1\right)$

$= \lim_{n \to +\infty} \left(5 - \dfrac{3}{n}\right) = 5 - 0 = 5$

$\lim_{n \to +\infty} (f(v_n)) = \lim_{n \to +\infty} \left(\left(2 + \dfrac{1}{n}\right) - 2\right) = \lim_{n \to +\infty} \left(\dfrac{1}{n}\right) = 0$

$\therefore \lim_{n \to +\infty} (f(u_n)) \neq \lim_{n \to +\infty} (f(v_n))$

Therefore the limit of a function at a point $x = 2$ does not exist.

Find two sequences $\{u_n\}$ and $\{v_n\}$ such that:
- $u_n < 2$ and $v_n > 2$
- $\lim_{n \to +\infty} u_n = \lim_{n \to +\infty} v_n = a$

Use properties of limits of sequences to show that $\lim_{n \to +\infty} (f(u_n)) \neq \lim_{n \to +\infty} (f(v_n))$

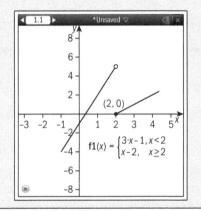

Example 9 illustrates the relation between a proof demonstrating that limit of a function does not exist, and the intuitive way you learned to find limits of functions as part of the core course. In practical terms, you could reach the same conclusion by substituting $x = 2$ in both branches of the piecewise function. However, this method of substitution does not give a rigorous result about the limit of a function in the way that the proof (shown in example 9) does, although it is sufficient for examination purposes.

Example 9 shows a function whose behavior to the left of $x = 2$ is different from its behavior to the right of $x = 2$. When we want to study the behavior of the function on just one side of the point $x = 2$ we use lateral limits:

Definition of right limit at a point:

Let I be an interval over the real numbers. Let $f : I \to \mathbb{R}$ and $a \in I$.

Then: $\lim_{x \to a^+} f(x)$ exists and $\lim_{x \to a^+} f(x) = b$ if and only if for any sequence $\{a_n\}$ such that

- $a_n \in I$ for all $n \in \mathbb{Z}^+$
- $a_n > a$ for all $n \in \mathbb{Z}^+$
- $\lim_{n \to \infty} a_n = a$

we have $\lim_{n \to \infty} (f(a_n)) = b$.

The $\lim_{x \to a^+} f(x)$ is called right limit of f at $x = a$.

The definition of left limit is similar:

Definition of left limit at a point:

Let I be an interval of real numbers. Let $f : I \to \mathbb{R}$ and $a \in I$.

$\lim_{x \to a^-} f(x)$ exists and $\lim_{x \to a^-} f(x) = b$ if and only if for any sequence $\{a_n\}$ such that

- $a_n \in I$ for all $n \in \mathbb{Z}^+$
- $a_n > a$ for all $n \in \mathbb{Z}^+$
- $\lim_{n \to \infty} a_n = a$

> Note that in these definitions of right and left limits, I can be any type of interval: open, closed, semi-open,..., etc.

we have $\lim_{n \to \infty} (f(a_n)) = b$.

When a is an endpoint of the interval I, we define the limit of the function at that point as a lateral limit. If a is not an endpoint of I, the $\lim_{x \to a} f(x)$ exists exactly when both lateral limits exist and $\lim_{x \to a^-} f(x) = \lim_{x \to a^+} f(x)$.

Example 10

Consider the function f defined by $f(x) = \begin{cases} \dfrac{1}{x-1}, & -2 < x < 0 \\ x-1, & 0 \leq x < 2 \end{cases}$

Show that **a** $\lim\limits_{x \to -2} f(x) = -\dfrac{1}{3}$ **b** $\lim\limits_{x \to 0} f(x) = -1$ **c** $\lim\limits_{x \to 2} f(x) = 1$.

a Let $\{u_n\}$ be any sequence such that $-2 < u_n < 0$ for all $n \in \mathbb{Z}^+$ and $\lim\limits_{n \to +\infty} u_n = -2$.

$$\lim_{n \to +\infty}(f(u_n)) = \lim_{n \to +\infty}\left(\dfrac{1}{u_n - 1}\right) = \dfrac{1}{\lim\limits_{x \to \infty} u_n - 1}$$

$$= \dfrac{1}{-2-1} = -\dfrac{1}{3}$$

$\therefore \lim\limits_{x \to -2} f(x) = \lim\limits_{x \to -2^+} f(x) = -\dfrac{1}{3}$

Use definition of right limit at $x = -2$ and properties of limits of sequences.

b Let $\{u_n\}$ be any sequence such that $-2 < u_n < 0$ for all $n \in \mathbb{Z}^+$ and $\lim\limits_{n \to +\infty} u_n = 0$.

$$\lim_{n \to +\infty}(f(u_n)) = \lim_{n \to +\infty}\left(\dfrac{1}{u_n - 1}\right) = \dfrac{1}{0-1} = -1$$

Let $\{v_n\}$ be any sequence such that $0 < v_n < 2$ for all $n \in \mathbb{N}^+$ and $\lim\limits_{n \to +\infty} v_n = 0$.

$$\lim_{n \to +\infty}(f(v_n)) = \lim_{n \to +\infty}(v_n - 1) = 0 - 1 = -1$$

$\therefore \lim\limits_{x \to 0^-} f(x) = \lim\limits_{x \to 0^+} f(x) = -1 \Rightarrow \lim\limits_{x \to 0} f(x) = -1$

Use definition of lateral limits at $x = 0$ and properties of limits of sequences to show that both limits exist and are equal to -1.

c Let $\{u_n\}$ be any sequence such that $0 < u_n < 2$ for all $n \in \mathbb{N}^+$ and $\lim\limits_{n \to +\infty} u_n = 2$.

$$\lim_{n \to +\infty}(f(u_n)) = \lim_{n \to +\infty}(u_n - 1) = 2 - 1 = 1$$

$\therefore \lim\limits_{x \to 2} f(x) = \lim\limits_{x \to 2^-} f(x) = 1$

Use definition of left limit at $x = 2$ and properties of limits of sequences.

Note that in the example above, the domain of the function is the open interval $I = \,]-2, 2[$. For the study of limits of functions at a point $x = a$, the value of the function at $x = a$ is not important, nor is whether or not the function is defined at $x = a$.

Heinrich Eduard Heine (1821–81) was a German mathematician best known for his work in real analysis, but he also investigated basic hypergeometric series. Heine studied first at the University of Berlin, and then went to the University of Göttingen where he attended lectures by Gauss. As a professor, Heine was greatly liked because his lectures were distinguished by clarity. He always pointed out to his students that it was not sufficient to read text books and manuals; rather it was necessary for them to study the approach to mathematics in the original papers. In this way, he promoted research among undergraduate students.

When the domain of the function contains intervals of the form $I =]a, +\infty[$ or $I =]-\infty, a[$, we can also use sequences to define $\lim_{x \to \pm\infty} f(x)$:

Definition: Let $f: D_f \to \mathbb{R}$ and $I =]a, +\infty[\subseteq D_f$. Then $\lim_{x \to +\infty} f(x)$ exists and $\lim_{x \to +\infty} f(x) = L$ if and only if for any sequence $\{a_n\}$ such that $a_n \in I$ for all $n \in \mathbb{Z}^+$ and $\lim_{n \to \infty} a_n = +\infty$, then $\lim_{n \to \infty}(f(a_n)) = L$.

> D_f is the domain of the function. This is either specified along with the function itself, or else can be taken as the largest possible subset of the real numbers on which the function can be defined.

Definition: Let $f: D_f \to \mathbb{R}$ and $I =]-\infty, a[\subseteq D_f$. $\lim_{x \to -\infty} f(x)$ exists and $\lim_{x \to -\infty} f(x) = L$ if and only if for any sequence $\{a_n\}$ such that $a_n \in I$ for all $n \in \mathbb{Z}^+$ and $\lim_{n \to \infty} a_n = +\infty$, $\lim_{n \to \infty}(f(a_n)) = L$.

> The limit L can either be a real number or $\pm\infty$.

Important consequences of these definitions:

1. To show that the $\lim_{x \to \pm\infty} f(x)$ does not exist, it is enough to find two sequences $\{u_n\}$ and $\{v_n\}$ such that $\lim_{n \to +\infty} u_n = \lim_{n \to +\infty} v_n = \pm\infty$ but $\lim_{n \to +\infty}(f(u_n)) \neq \lim_{n \to +\infty}(f(v_n))$.

2. All the results about algebra of limits, and how to deal with indeterminate forms which we studied in the previous section, can also be applied to limits of functions.

> The notation I represents any interval within the real numbers. The interval I can be a closed interval $[a,b]$ an open interval $]a, b[$, a semi-open interval $[a,b[$, $[a,\infty[$, $]a, b]$ or $]-\infty, b]$, or even the whole set of real numbers $]-\infty, +\infty[$.

Example 11

Given the function f, defined by $f(x) = \sin(x)$, show that $\lim\limits_{x \to +\infty} f(x)$ does not exist.

Let $u_n = \pi n$ and $v_n = 2\pi n + \dfrac{\pi}{2}$.

$\lim\limits_{n \to +\infty} u_n = \lim\limits_{n \to +\infty} v_n = +\infty$

$\lim\limits_{n \to +\infty} \left(f(u_n) \right) = \lim\limits_{n \to +\infty} \left(\sin(\pi n) \right) = \lim\limits_{n \to +\infty} (0) = 0$

$\lim\limits_{n \to +\infty} \left(f(v_n) \right) = \lim\limits_{n \to +\infty} \left(\sin\left(2\pi n + \dfrac{\pi}{2} \right) \right) = \lim\limits_{n \to +\infty} (1) = 1$

$\therefore \lim\limits_{n \to +\infty} \left(f(u_n) \right) \neq \lim\limits_{n \to +\infty} \left(f(v_n) \right)$

Therefore $\lim\limits_{x \to +\infty} f(x)$ does not exist.

Graph the function and select special sequences of points on its graph. For example the sequence of the zeros or maximum points, i.e. two sequences $\{u_n\}$ and $\{v_n\}$ such that $\lim\limits_{n \to +\infty} (f(u_n)) \neq \lim\limits_{n \to +\infty} (f(v_n))$.

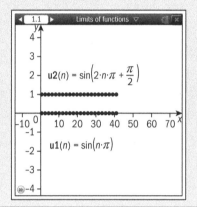

Example 12

Calculate the following limits:

a $\lim\limits_{x \to +\infty} \dfrac{3x+1}{x^2 + 2x + 1}$ **b** $\lim\limits_{x \to +\infty} \left(\sqrt{x+1} - \sqrt{x} \right)$

a $\lim\limits_{x \to +\infty} \dfrac{3x+1}{x^2 + 2x + 1} = \dfrac{\infty}{\infty}$ (ind)

Identify indeterminate form.

$\lim\limits_{x \to +\infty} \dfrac{3x+1}{x^2 + 2x + 1} = \lim\limits_{x \to +\infty} \dfrac{\dfrac{3}{x} + \dfrac{1}{x^2}}{1 + \dfrac{2}{x} + \dfrac{1}{x^2}} = \dfrac{0}{1} = 0$

Simplify the expression and then apply Theorem 2.

b $\lim\limits_{x \to +\infty} \left(\sqrt{x+1} - \sqrt{x} \right) = +\infty - \infty$ (ind)

Identify indeterminate form.

$\lim\limits_{x \to +\infty} \left(\sqrt{x+1} - \sqrt{x} \right) = \lim\limits_{x \to +\infty} \dfrac{\left(\sqrt{x+1} - \sqrt{x} \right)\left(\sqrt{x+1} + \sqrt{x} \right)}{\sqrt{x+1} + \sqrt{x}}$

Simplify the expression using difference of squares.

$\lim\limits_{x \to +\infty} \dfrac{x+1-x}{\sqrt{x+1} + \sqrt{x}} = \lim\limits_{x \to +\infty} \dfrac{1}{\sqrt{x+1} + \sqrt{x}} = \dfrac{1}{\infty} = 0$

Then apply Theorem 2.

Exercise 1D

1 Show that the function f defined by $f(x) = \begin{cases} x^2 - 1, & x < 1 \\ 3x - 1, & x \geq 1 \end{cases}$ has no limit at $x = 1$.

Use technology to confirm the result graphically. Sketch the graph of f.

2 Sketch the graph of each of the following functions, and decide whether or not the function has a limit at the given point. Justify your answers algebraically.

a $f(x) = \begin{cases} \sin(\pi x), & x < 1 \\ \cos\left(\dfrac{\pi x}{2}\right), & x \geq 1 \end{cases}$ at $x = 1$;

b $g(x) = \begin{cases} e^x - 1, & x < 0 \\ \ln(x), & x > 0 \end{cases}$ at $x = 0$;

c $f(x) = \begin{cases} \dfrac{3x+1}{x+1}, & x < -2 \\ \dfrac{x+1}{3x+1}, & x \geq -2 \end{cases}$ at $x = -2$.

3 Calculate, if possible, the following limits:

a $\lim\limits_{x \to +\infty} \dfrac{5x - 1}{2x^2 + 3x - 1}$

b $\lim\limits_{x \to +\infty} \dfrac{5x^2 - x + 1}{x^2 - 3x + 1}$

c $\lim\limits_{x \to +\infty} \left(\sqrt{2x+1} - \sqrt{x+1}\right)$

4 Given the function f defined by $f(x) = \cos(\pi x)$,

a Sketch the graph of f.

b Show that $\lim\limits_{x \to +\infty} f(x)$ does not exist.

Review exercise

1 Determine the limit of each of the following sequences:

   a $u_n = \dfrac{n^3}{2n^3 - 1}$

   b $u_n = \dfrac{n^3}{2^n}$

   c $u_n = \left(\dfrac{2}{3}\right)^n$

   d $u_n = \sqrt{n+3} - \sqrt{n}$

   e $u_n = \dfrac{1 + 2 + 3 + \ldots + 2n}{n^2}$

   f $u_n = \dfrac{\sum_{k=1}^{n} 2 \cdot 3^k}{5^n}$

2 Give an example to **disprove** the statement: '*If $a_n < 0$ for all $n \in \mathbb{Z}^+$ then* $\lim\limits_{n \to +\infty} a_n < 0$'.

EXAM-STYLE QUESTIONS

3 Use the squeeze theorem to find the value of $\lim\limits_{n \to \infty} \left(\sum_{k=0}^{n} \dfrac{3}{n^2 + k} \right)$

4 Use the squeeze theorem to prove that the sequence $u_n = \sum_{k=1}^{n} \dfrac{n + \cos(k\pi)}{4n^2 + 3}$ converges.

5 Consider the sequence $\dfrac{1}{2 \times 3}, \dfrac{1}{5 + 6}, \dfrac{1}{8 \times 9}, \ldots$

   a Write down an expression for the general term u_n of the sequence.

   b Hence, find $\lim\limits_{n \to \infty} \dfrac{u_n}{u_{n+1}}$

   c Show that $(\sin(n) \cdot u_n)^n$ is convergent, stating its limit.

6 Show that $\sqrt{n} \leq n!$ for $n \in \mathbb{Z}^+$. Hence, find the value of $\lim\limits_{n \to \infty} \dfrac{n! + \sqrt{n}}{(n+2)!}$

7 Consider the sequence defined by the following recurrence formula:
$$\begin{cases} a_1 = 3 \\ a_{n+1} = \dfrac{a_n + 1}{2}, n \in \mathbb{Z}^+ \end{cases}$$

   a Show that $a_n > 0$ for all $n \in \mathbb{Z}^+$

   b Prove, using the method of mathematical induction, that $a_{n+1} - a_n < 0$ for all $n \in \mathbb{Z}^+$

   c Justify that $\{a_n\}$ is convergent and then determine its limit.

   d Determine $\lim\limits_{n \to \infty} (a_n)^n$.

Chapter 1 summary

Convergent sequences

$\{u_n\}$ is a convergent sequence with $\lim_{n\to\infty} u_n = L$ if and only if for any $\varepsilon > 0$ there exists a minimum order $m \in \mathbb{Z}^+$ such that, for all $n \geq m \Rightarrow |u_n - L| < \varepsilon$.

Theorems about convergence of subsequences

- If $\{b_n\} \subseteq \{a_n\}$ a subsequence of a convergent sequence $\{a_n\}$, then $\{b_n\}$ is also a convergent sequence and $\lim_{n\to\infty} b_n = \lim_{n\to\infty} a_n$.
- If $\{b_n\} \subseteq \{a_n\}$ and $\{c_n\} \subseteq \{a_n\}$ are subsequences of a sequence $\{a_n\}$ and $\lim_{n\to\infty} b_n \neq \lim_{n\to\infty} c_n$ then $\{a_n\}$ is a divergent sequence

Squeeze Theorem

Consider three sequences such that:

- There exists some $p \in \mathbb{Z}^+$ such that $u_n \leq v_n \leq w_n$ for all $n \geq p \in \mathbb{Z}^+$
- $\{u_n\}$ and $\{w_n\}$ converge and $\lim_{n\to\infty} u_n = \lim_{n\to\infty} w_n = L$

Then $\{v_n\}$ converges and $\lim_{n\to\infty} v_n = L$

Theorems about limits of sequences: algebra of limits

Let $\{u_n\}$ and $\{v_n\}$ be convergent sequences and $\lim_{n\to\infty} u_n = L_1$ and $\lim_{n\to\infty} v_n = L_2$. Then:

$\lim_{n\to\infty}(u_n + k) = L_1 + k$ $\quad$ $\lim_{n\to\infty}(ku_n) = kL_1$, for any $k \in \mathbb{R}$

$\lim_{n\to\infty}(-u_n) = -L_1$ $\quad$ $\lim_{n\to\infty}\dfrac{1}{u_n} = \dfrac{1}{L_1}$ when $L_1 \neq 0$

$\lim_{n\to\infty}(u_n + v_n) = L_1 + L_2$ $\quad$ $\lim_{n\to\infty}(u_n - v_n) = L_1 - L_2$

$\lim_{n\to\infty}(u_n \cdot v_n) = L_1 L_2$ $\quad$ $\lim_{n\to\infty}\left(\dfrac{u_n}{v_n}\right) = \dfrac{L_1}{L_2}$ when $L_2 \neq 0$.

$\lim_{n\to\infty}|u_n| = +\infty \Rightarrow \lim_{n\to\infty}\dfrac{1}{u_n} = 0$ $\quad$ $\lim_{n\to\infty} u_n = 0 \Rightarrow \lim_{n\to\infty}\left|\dfrac{1}{u_n}\right| = +\infty$

Algebra of infinity

$(\pm\infty) + (\pm\infty) = \pm\infty$	$(\pm\infty) \times (\pm\infty) = +\infty$	$a + (\pm\infty) = \pm\infty, a \in \mathbb{R}$
$(\pm\infty) - (\pm\infty) =$ indeterminate	$(\pm\infty) \times (\mp\infty) = \mp\infty$	$a - (\pm\infty) = \mp\infty, a \in \mathbb{R}$
$a \times (\pm\infty) = \pm\infty, a > 0$	$a \times (\pm\infty) = \mp\infty, a < 0$	$\infty^n = \infty, n \in \mathbb{Z}^+$
$\dfrac{a}{\infty} = 0, a \in \mathbb{R}$	$\dfrac{0}{0} =$ indeterminate	$\dfrac{\infty}{\infty} =$ indeterminate
$\dfrac{\infty}{a} = \infty, a \in \mathbb{R}, a \neq 0$	$\sqrt[n]{+\infty} = +\infty, n$ even $\sqrt[n]{\pm\infty} = \pm\infty, n$ odd	$0 \times \infty =$ indeterminate

2 Smoothness in mathematics

CHAPTER OBJECTIVES:
9.3 Continuity and differentiability of a function at a point. Continuous and differentiable functions.
9.6 Rolle's Theorem and the Mean Value Theorem.
9.7 L'Hôpital rule and evaluation of limits. Applications to limits of sequences.

Before you start

You should know how to:

1 Use a GDC to graph polynomial, rational, exponential, logarithmic, and trigonometric functions; and compositions of these. Recognise distinguishing features of the graphs of these functions.
 e.g. graph $f(x) = \dfrac{\sin(x)}{x^2 - 1}, x \neq \pm 1$
 Axes intercept $(k\pi, 0), k \in \mathbb{Z}$
 Asymptotes: $x = \pm 1$ and $y = 0$;
 No maximum or minimum points.

2 Use a GDC to graph piecewise functions and recognise distinguishing features of their graphs.
 e.g. graph $f(x) = \begin{cases} 1 - x^2, & x \leq 1 \\ 4 - x, & x > 1 \end{cases}$
 Axes intercept $(1, 0), (-1, 0), (0, 1), (4, 0)$;
 No asymptotes;
 Local maximum $f(x) = 1$ at $x = 0$;
 Local minimum $f(x) = 0$ at $x = 1$.

3 Use properties of limits of functions.
 e.g. find
 $\lim\limits_{x \to 0} \left(\dfrac{3\sin(x)}{2x} - e^{2x} \right) = \dfrac{3}{2} \lim\limits_{x \to 0} \dfrac{\sin(x)}{x} - \lim\limits_{x \to 0} e^{2x}$
 $= \dfrac{3}{2} \cdot 1 - 1 = \dfrac{1}{2}$

Skills check:

1 Graph the following functions, showing clearly the axes intercepts and any maxima and minima points. Also write down the equations of any asymptotes:
 a $f(x) = \dfrac{x}{x - 1}, x \neq 1$ **b** $f(x) = xe^{x-1} + 1$
 c $f(x) = \sin\left(2x - \dfrac{\pi}{3}\right) + 1$ **d** $f(x) = \ln(2x - 1)$

2 Graph the following functions, showing clearly the axes intercepts and any maxima and minima points. Also write down the equations of any asymptotes:
 a $f(x) = \begin{cases} 1 - x, & x \leq 2 \\ (x - 2)^2 - 1, & x > 2 \end{cases}$
 b $f(x) = \begin{cases} |x - 2|, & x \leq 0 \\ x^2 - x - 2, & x > 0 \end{cases}$
 c $f(x) = \begin{cases} 2^{x-1}, & x \leq 2 \\ 4 - x, & x > 2 \end{cases}$

3 Use properties of limits to find the following limits:
 a $\lim\limits_{x \to 1}\left((4x - 1)^2 - 5\right)$ **b** $\lim\limits_{x \to 0}\left(\dfrac{\tan(2x)}{2x} - 3\right)^3$

Exploring continuous and differentiable functions

"Logic sometimes makes monsters. For half a century we have seen a mass of bizarre functions which appear to be forced to resemble as little as possible honest functions which serve some purpose. More of continuity, or less of continuity, more derivatives, and so forth. Indeed, from the point of view of logic, these strange functions are the most general; on the other hand those which one meets without searching for them, and which follow simple laws appear as a particular case which does not amount to more than a small corner. In former times when one invented a new function it was for a practical purpose; today one invents them purposely to show up defects in the reasoning of our fathers and one will deduce from them only that."

—**Henri Poincare on Pathological functions, 1899**

In the 18th and 19th centuries, many mathematicians were producing results about functions, but soon after were being challenged with counter-examples which contradicted their results. All these counter-example functions shared a common element: they were not smooth enough to obey the result that had been previously proposed. Thus, mathematicians realized that they had to restrict their results about functions by saying that the result would only hold true for 'smooth functions'.

But can smoothness be defined mathematically? Many mathematicians facing this problem complained that the hypotheses required to avoid what they called 'pathological functions' (functions that were continuous, but no where differentiable) spoilt the elegance of classical analysis. Some held up their hands in genuine horror at the dreadful plague of 'pathological functions', with which nothing mathematical could be done.

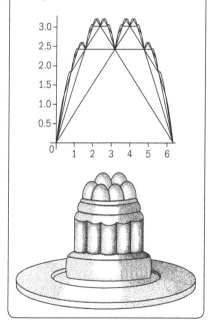

The Blancmange function is a modern example of a pathological function that is continuous everywhere but nowhere differentiable. Its name derives from the resemblance to the shape of the dessert commonly made with milk or cream and sugar, and thickened with gelatin.

2.1 Continuity and differentiability on an interval

Between the 18th and 19th centuries, famous mathematicians like Gauss and Cauchy made the first attempts to create an algebra for calculus and define precisely the concepts of *limit of a function*, *continuity*, and *differentiability at a point*. Cauchy, for example, defined *limit of a function* precisely, using very similar terms to the ones we use today. He was the first to interpret $\lim_{x \to a} f(x) = b$ as a relation between two variables: when the difference $\Delta x = x - a$ becomes infinitely small, the difference $\Delta y = f(x) - b$ also becomes infinitely small.

Given a function f it is not necessarily true that $\lim_{x \to a} f(x) = f(a)$.

For example, for

$$f(x) = \begin{cases} 2, & x = 3 \\ (x-2)^2 - 1, & x \neq 3 \end{cases}$$

we see that $\lim_{x \to 3} f(x) = 0 \neq 2 = f(3)$.

Therefore it makes sense to distinguish functions for which $\lim_{x \to a} f(x) = f(a)$. We call them continuous functions.

> Augustin-Louis Cauchy (1789–1857), was regarded as one of the two foremost mathematicians of his time, the other being Gauss. In 1805, Cauchy began studying at the École Polytechinique, one of the famous schools founded by Napoleon to train engineers for the French Army, and afterward he worked as an engineer on bridges and railways. Cauchy published prolifically and by the time he returned to Paris in 1813 had already published several papers in geometry, number theory and determinants. Cauchy's *Cours d'Analys de l'École Polytechinque*, written in 1821, had a major impact on today's understanding of limits, continuity and integrals. In fact, Cauchy gave the first reasonably rigorous foundation for calculus. He defined derivatives as the limit of the difference quotient. Defining the definite integral, and not the antiderivative, enabled him to integrate non-continuous functions as we do. He proved the (Cauchy) Mean Value Theorem via Rolle's Theorem and used it to prove the Fundamental Theorem of Calculus.

Definition: Suppose $f : D_f \to \mathbb{R}$. Given $a \in D_f$, if $\lim_{x \to a} f(x) = f(a)$ then f is **continuous** at the point a. This means that:

1 $\lim_{x \to a^-} f(x)$ and $\lim_{x \to a^+} f(x)$ exist, and $\lim_{x \to a^-} f(x) = \lim_{x \to a^+} f(x)$

2 $\lim_{x \to a} f(x) = f(a)$

If f is continuous at each point on an interval $I \subseteq D_f$, we say that the function is continuous on I.

Most functions you have studied as part of the Math HL core course are continuous on their domains: their graphs contain no breaks or jumps. These are the graphs that, for each interval of the domain, can be drawn in one go, without lifting the pencil or pen from the paper.

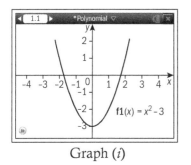 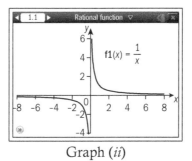 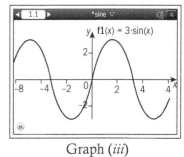

Graph (*i*)   Graph (*ii*)   Graph (*iii*)

In the examples above, graphs (i) and (iii) can be drawn in one go without lifting the pencil from the paper because these functions are continuous over $\mathbb{R}$. Graph (ii) has two branches which correspond to the two intervals of the domain of the function: $]-\infty, 0[$ and $]0, +\infty[$. Both of these branches can be drawn separately without lifting the pencil.

However, we can also find examples of functions that are not continuous at one or more points on their domain:

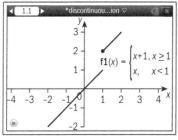 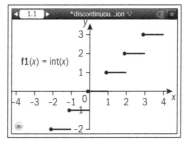 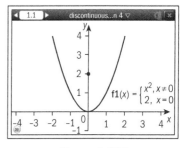

Example (*i*)   Example (*ii*)   Example (*iii*)

In general a function is not continuous at a point $x = a$ when:

1 $\lim\limits_{x \to a^-} f(x) \neq \lim\limits_{x \to a^+} f(x)$ and in this case we say that $\lim\limits_{x \to a} f(x)$ does not exist;

2 $\lim\limits_{x \to a^-} f(x) = \lim\limits_{x \to a^+} f(x)$ (i.e. $\lim\limits_{x \to a} f(x)$ exists) but $\lim\limits_{x \to a} f(x) \neq f(a)$.

The examples (i) and (ii) show graphs of functions that fail condition (1) while example (iii) shows a graph of a function that fails condition (2). Unfortunately the graph displayed by the GDC does not show that there is hole in the graph and we need to rely on the expression to verify that $\lim\limits_{x \to 0} x^2 = 0 \neq 2 = f(0)$.

Example 1

Use a GDC to graph the function defined by $f(x) = \begin{cases} 3x-1, & x>1 \\ x^2+1, & x \leq 1 \end{cases}$

Show that f is continuous at $x = 1$.

$\lim_{x \to 1^-}(x^2+1) = 1 + 1 = 2$

$\lim_{x \to 1^+}(3x-1) = 3 \times 1 - 1 = 2$

$\therefore \lim_{x \to 1} f(x) = 2 = f(1)$

$\therefore f$ is continuous at $x = 1$.

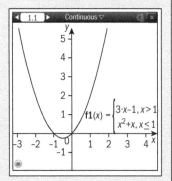

Observe that the graph shows no gaps.

Calculate the left and right limits of f at $x = 1$ and show that

$\lim_{x \to 1^-} f(x) = \lim_{x \to 1^+} f(x) = f(1)$

Continuity conditions (1) and (2) allow us to test continuity for most common functions. The following examples show you that in the case of pathological functions we need to consider whether $\lim_{x \to a^-} f(x)$ and $\lim_{x \to a^+} f(x)$ even exist.

Example 2

Use a GDC to graph the function defined by $f(x) = \begin{cases} \sin\left(\dfrac{1}{x}\right), & x \neq 0 \\ 0, & x = 0 \end{cases}$

Show that f is not continuous at $x = 0$.

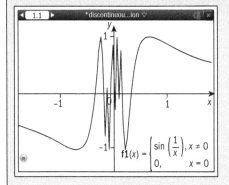

Consider the sequences $u_n = \dfrac{1}{\pi n}$ and $v_n = \dfrac{1}{2\pi n + \dfrac{\pi}{2}}, n \in \mathbb{Z}^+$

$\lim_{n \to \infty} f(u_n) = 0 \neq \lim_{n \to \infty} f(v_n) = 1$.

So, $\lim_{x \to 0} f(x)$ does not exist.

Observe that the graph oscillates wildly as $x \to 0$.

For example, there are an infinite number of zeros and maximum points on the interval $]0,1[$ that allow us to find the expressions of the sequences u_n and v_n.

Use the definition of limit studied in Chapter 1: if two sequences $\{u_n\}$ and $\{v_n\}$ converge and $\lim_{n \to \infty} u_n = \lim_{n \to \infty} v_n = a$, but $\lim_{n \to \infty} f(u_n) \neq \lim_{n \to \infty} f(v_n)$ then $f(x)$ has no limit at a.

In the example above, the function was continuous everywhere apart from at $x = 0$.

The example below shows you an extreme case: a function that is not continuous at any point of its domain. Although the expression of the function is simple, it is difficult to imagine its graph, since close to any irrational number there are always infinitely many rational numbers and close to any rational number there are always infinitely many irrational numbers.

Example 3

Show that the function $f(x) = \begin{cases} 1, & x \text{ irrational} \\ 0, & x \text{ rational} \end{cases}$ is not continuous at any point $x = a$.	
Let $a \in \mathbb{R}$. Consider the sequences $\{u_n\}$ and $\{v_n\}$ such that $\lim_{n \to \infty} u_n = a$ and $u_n \in \mathbb{Q} \setminus \{a\}$ and $\lim_{n \to \infty} v_n = a$ and $v_n \notin \mathbb{Q}$, $v_n \neq a$. Then $\lim_{n \to \infty} f(u_n) = 0 \neq \lim_{n \to \infty} f(v_n) = 1$. So, $\lim_{x \to a} f(x)$ does not exist.	Use definition of limit studied in **1.4**: 'if two sequences $\{u_n\}$ and $\{v_n\}$ converge to the same limit, but the two sequences $\{f(u_n)\}$ and $\{f(v_n)\}$ converge to different limits, then the function f has no limit at a.' Note: $\mathbb{Q} \setminus \{a\}$ means all rational numbers, except a.

Exercise 2A

1 Use a GDC to graph the following functions and state whether or not they have a point on their domain where they are discontinuous.

a $f(x) = \begin{cases} x^2 - 3x + 1, & x \leq 1 \\ 4 - x, & x > 1 \end{cases}$

b $g(x) = \begin{cases} e^{-x^2}, & x > 1 \\ 1 - e^x, & x \leq 1 \end{cases}$

c $h(x) = \begin{cases} x e^{-x}, & x \neq 0 \\ 0, & x = 0 \end{cases}$

d $i(x) = \begin{cases} x \ln\left|\frac{1}{x}\right|, & x \neq 0 \\ 0, & x = 0 \end{cases}$

2 Show that the function defined by $g(x) = \begin{cases} \cos\left(\frac{1}{x}\right), & x \neq 0 \\ 0, & x = 0 \end{cases}$ is not continuous at $x = 0$.

3 Show that the function defined by $f(x) = \begin{cases} x \sin\left(\frac{1}{x}\right), & x \neq 0 \\ 0, & x = 0 \end{cases}$ is continuous at $x = 0$.

4 Use properties of limits of functions to show that if f and g are continuous functions at a point $x = a$, then:

a $f + g$ is continuous at $x = a$;

b $f - g$ is continuous at $x = a$;

c $f \cdot g$ is continuous at $x = a$;

d $\dfrac{f}{g}$ is continuous at $x = a$ when $g(a) \neq 0$.

Investigation – Composition of continuous functions

Consider the functions $f(x) = \begin{cases} x+1, & x \neq 0 \\ 0, & x = 0 \end{cases}$ and $g(x) = \begin{cases} x, & x \neq 1 \\ 0, & x = 1 \end{cases}$.

a Show that f is discontinuous at $x = 0$ and g is discontinuous at $x = 1$ but $f \circ g$ is continuous at $x = 0$ and $g \circ f$ is continuous at $x = 1$.

Consider $h(x) = \begin{cases} x-1, & x \leq 2 \\ x^2 - 1, & x > 2 \end{cases}$

b Show that h and f are continuous at $x = 1$ but $h \circ f$ is discontinuous at $x = 1$.

c Investigate further and establish conditions for which $f \circ g$ is continuous at $x = a$.

2.2 Theorems about continuous functions

The most interesting and useful results about continuous functions occur for functions that are actually continuous at all points on their domain. For the rest of chapter we are going to focus on these type functions, and consider discontinuities as exceptions that may occur at just a few points on the domain of a function.

The following list of results summarizes familiar properties of continuous functions that make them good tools to work with:

Theorem 1: Let f and g be continuous functions on their domains, D_f and D_g, respectively. Then:

i $f \pm g$ and $f \cdot g$ are continuous on $D = D_f \cap D_g$;

ii $\dfrac{f}{g}$ is continuous on $D = \{x \in D_f \cap D_g | g(x) \neq 0\}$;

iii f^n is continuous on D_f, $n \in \mathbb{Z}^+$;

iv $\sqrt[n]{f}$ is continuous on D_f when $n \in \mathbb{Z}^+$ is odd, and on $D_f \cap \{x | f(x) \geq 0\}$ when n is even;

v $f \circ g$ is continuous on $D = \{x \in D_g | g(x) \in D_f\}$.

> As the proofs of these results are beyond the syllabus requirements, they will be omitted.

Theorem 2: Let f be a continuous one-one function on its domain. Then f^{-1} is also continuous on its domain.

The properties of continuous functions are even more useful when their domains are intervals, or if we restrict their study to an interval I.

Theorem 3: Let f be a continuous one-one function on $I \subset D_f$. Then both f and f^{-1} are either increasing or decreasing on I.

> The contrapositive of Theorem 3 gives us an important result:
> Suppose at least one of f or f^{-1} are neither increasing or decreasing on $I \subseteq D_f$. Then f is not a continuous one-one function on I.

Example 4

Use the contrapositive of Theorem 3 to show that the function $g(x) = x^2$ cannot be the inverse of $f(x) = \sqrt{x}$ when these functions are defined on their largest possible domains. State the inverse of f and whether it is an increasing or decreasing function.

The largest possible domain of g is $\mathbb{R}$. Let $I = \mathbb{R}$.
Now g is neither increasing nor decreasing on I.

By Theorem 3, g cannot be a continuous one-one function on I. Since g is not one-one on I, it cannot have an inverse function on I.

Hence, f is not the inverse of g on I.

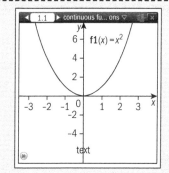

The inverse of f is defined by $f^{-1}(x) = x^2$, $x \geq 0$

Now consider the restriction of g to $I = [0, \infty[$.

Then $g(x) = \{x^2, x \geq 0\} = f^{-1}(x)$.

Both f and f^{-1} are increasing functions on the domain I.

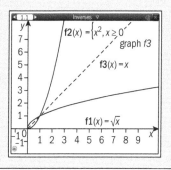

When the interval I is closed (e.g. $I = [a, b]$ for $a, b \in \mathbb{R}$, $a < b$), the results become even more powerful: continuous functions defined on closed intervals can be graphed inside a rectangle, such as the screen of your calculator. This means that we can actually see all the graph of these functions.

Example 5

Let $g(x) = \sin(\pi x^2)$, $x \in [-1, 1]$. Find the range of g and the values of x for which g has a maximum or a minimum.

From GDC, the range $g([-1, 1]) = [0, 1]$.

The absolute maximum occurs at $x = \pm\sqrt{\dfrac{1}{2}} = \pm 0.707$

(3 s.f.) and the absolute minimum at $x = 0$, and $x = \pm 1$.

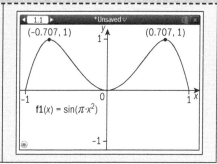

IB examinations usually set questions on open intervals, such as $I =]a, b[$. On open intervals, local maxima and minima are always zeros of the derivative function. However, the example above shows that when the interval is closed, the minima points can also occur at the endpoints of the domain, even if the derivative is not zero (or sometimes not even defined) at these endpoints.

Chapter 2

In example 5 the domain of the function is a closed interval: $[-1, 1]$. The image of the domain $f([-1, 1])$ is also a closed interval: $[0, 1]$. This occurs for continuous functions defined on closed intervals as stated in the following theorem:

Theorem 4: Let f be continuous on $[a, b] \subset D_f$, $a < b$.

Then $f([a, b]) = [c, d]$, $c < d$.

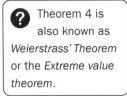

Theorem 4 is also known as Weierstrass' Theorem or the *Extreme value theorem*.

Look back at Example 5. It shows you that the range $f([a, b])$ is not necessarily $[f(a), f(b)]$ or $[f(b), f(a)]$. This result is only guaranteed when the function f is either increasing or decreasing on $[a, b]$, as the following example illustrates:

Example 6

Let $f(x) = 3 - x$, $x \in [0, 2]$. Find the range of f.

f is a decreasing function on $[0, 2]$.

$f(0) = 3$ is an absolute maximum.

$f(2) = 1$ is an absolute minimum.

$\therefore f([0, 2]) = [f(2), f(0)] = [1, 3]$

Since the function is decreasing, we applied [f(b), f(a)]

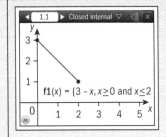

Read the range on the y-axis.

Sometimes it is also useful to look at the values of a function at the end points of an interval to decide whether or not equations have solutions on a given interval.

Example 7

Decide whether the equation $2e^{-x^2} = 1$ has solution(s) on the intervals $[0, 1]$ and $[-1, 0]$. State, with reasons, the number of solutions on each interval.

Let $f(x) = 2e^{-x^2}$.

$f(0) = 2 > 1$ and $f(1) = 2e^{-1} = \dfrac{2}{e} < 1$.

As the function is continuous on $[0, 1]$, its graph must cross the line $y = 1$ for $x \in [0, 1]$.

As the function is decreasing for some $x \geq 0$, $2e^{-x^2} = 1$ has exactly one solution on $[0, 1]$.

As the function is even, it is symmetrical about the y-axis, so $2e^{-x^2} = 1$ also has exactly one solution on $[-1, 0]$.

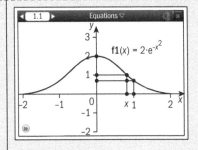

$f'(x) = -4xe^{-x^2} < 0$, for $x \in [0, 1]$, so f is decreasing on $[0, 1]$.

f even $\Rightarrow f(-x) = f(x)$

Theorem 5: Let f be continuous on $[a,b] \subset D_f$. If $f(a) < c < f(b)$ or $f(b) < c < f(a)$, then the equation $f(x) = c$ has at least one solution on $[a, b]$.

> 🔍 Theorem 5 is also known as Bolzano's theorem.

Note that if f is one-to-one, this value of c is unique on $[a, b]$.

Example 8

Let $f(x) = x^2$, $x \in [1, 2]$. Use Theorem 5 to show that $\sqrt{2}$ exists and $1 \leq \sqrt{2} \leq 2$.

f is continuous on $[1, 2]$. Let $c = 2$. Then $1 = f(1) < 2 < f(2) = 4$.	First, we need to show that $f(x) = x^2$ is continuous for $x \in [1, 2]$. Then we can apply Theorem 5.
Then Theorem 5 $\Rightarrow f(x) = 2$ has solutions on $[1, 2]$. So, there exists $x \in [1, 2]$ such that $f(x) = 2$, and thus $\sqrt{2}$ exists.	
Now f is increasing on $[1, 2]$, so $f([1, 2]) = [f(1), f(2)]$.	Applying Theorem 4 to a continuous, increasing function.
Thus, $f(1) = 1 < \sqrt{2} < 2 = f(2)$.	
Also, $(\sqrt{2})^2 = 2 \Rightarrow \sqrt{2} \leq 2$.	
Therefore $1 \leq \sqrt{2} \leq 2$.	

The following corollary formalizes a result that we have been using to solve equations: if the graph of a continuous function moves from below to above the graph of another continuous function on the same closed interval, these graphs must intersect at least once on this interval.

Corollary of Theorem 5: Let f and g be continuous on $[a, b]$. If $f(a) < g(a)$ and $g(b) < f(b)$, then the equation $f(x) = g(x)$ has at least one solution on $[a, b]$.

The proof is left to the student as an exercise.

Example 9

Use a GDC to graph the function defined by the expressions on the LHS and the RHS of arctan $(x) = x$.
Hence, show that the equation has at least one solution, stating clearly the theorem that guarantees its existence on the closed intervals you used.

Let $f(x) = \arctan(x) + 1$ and $g(x) = x$.

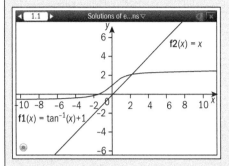

$f(0) = 1 > g(0) = 0$

$f(3) = \arctan(3) + 1 < g(3) = 3$

f and g are continuous on $\mathbb{R}$.

Inspect the graph and select a suitable interval to apply corollary of Theorem 5.

For example, the interval $[0, 3]$.

$\arctan(3) = 1.249...$

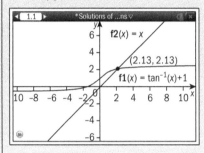

∴ By the corollary of Theorem 5, the graphs of f and g cross at a point between 0 and 3 and the equation has a solution on $]0, 3[$.

You can use your GDC to find an approximation of the solution.

Exercise 2B

1. Use a GDC to graph the functions defined by the expressions on the LHS and the RHS of the following equations. Hence, show that each equation has at least a solution, stating clearly the theorems that guarantee their existence and the closed intervals you used.

   a $(x + 1)e^x = 2$ **b** $x \ln(x) = 2 - 3x$ **c** $e^{x^2} = x + 2$

2. Find all continuous functions that satisfy the condition $(f(x))^2 = x^2$. Sketch their graphs and give reasons that support your answer.

3. Let f be a continuous function defined on $[a, b]$. Show that if $f(x) \in \mathbb{Q}$ for all $x \in [a, b]$ then f is a constant function. Hence state all possible values for c if $f(x) = c$.

EXTENSION QUESTION

4. Show that any polynomial of odd degree with real coefficients has at least one zero in $\mathbb{R}$. Explain your reasoning in detail and state the theorems you used.

2.3 Differentiable functions: Rolle's Theorem and Mean Value Theorem

In the previous section, we looked at continuous functions. These are well-behaved functions whose graphs have no breaks or jumps. We listed a collection of useful theorems that characterise their behaviour.

In this section we are going to establish important results about an even more specific class of functions that are all continuous, but are also differentiable, and whose graphs not only have no breaks or jumps but also have no bends or wild oscillations!

This means that we will exclude from our study hereafter functions like the ones below:

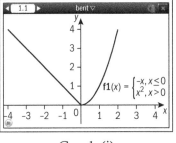 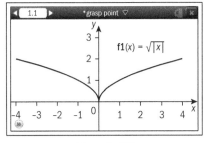 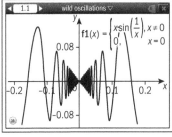

Graph (i)   Graph (ii)   Graph (iii)

In the core course you learnt that the tangent to a curve at a point $x = a$ can be seen as the limit case of secants to the curve through the points $(a, f(a))$ and $(a + h, f(a + h))$.

If the tangent to the curve exists at $x = a$ then its gradient (or slope) m is given by

$$m = \lim_{h \to 0} \frac{f(a+h) - f(a)}{h}.$$

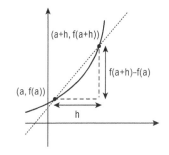

The graphs (i) – (iii) show functions that have no tangent line defined at $x = 0$. The following example shows you how to prove it for graph (i).

Example 10

Show that the graph of the function defined by $f(x) = \begin{cases} -x, & x < 0 \\ x^2, & x \geq 0 \end{cases}$ does not have a tangent at $x = 0$.

$\lim_{h \to 0^-} \frac{f(h) - f(0)}{h} = \lim_{h \to 0^-} \frac{-h - 0}{h} = -1;$

$\lim_{h \to 0^+} \frac{f(h) - f(0)}{h} = \lim_{h \to 0^+} \frac{h^2 - 0}{h} = \lim_{h \to 0^+} h = 0.$

Therefore $\lim_{h \to 0} \frac{f(a+h) - f(a)}{h}$ does not exist.

Use definition of gradient of tangent at the point $x = 0$ and show that

$\lim_{h \to 0} \frac{f(a+h) - f(a)}{h}$ does not exist

(calculate the left and right limit and show that they are not equal)

The following definition tells you that the tangent at a point $x = a$ is only defined when the function is differentiable at $x = a$.

Definition: A continuous function f is differentiable at $x = a$ if and only if $\lim_{h \to 0} \dfrac{f(a+h) - f(a)}{h}$ exists and is finite. In this case we represent this limit by $f'(a)$ and call it the derivative of f at $x = a$.

If a function is differentiable at all points on an interval I of its domain we say that the function is differentiable on I and we define the derivative function of f as

$$f'(x) = \lim_{h \to 0} \frac{f(x+h) - f(x)}{h}, \ x \in I.$$

Graphs of differentiable functions are locally linear because we can draw the tangent to the graph at any point. Most functions we have covered so far are differentiable at most points of their domains. In general you just need to analyze in detail the behaviour of functions at a small number of points. If you graph the function you should be able to spot these critical points: they are usually end points of the intervals of the domain of the function.

The following examples show that a function can be continuous on its domain but not differentiable at least at one point; either because $\lim_{h \to 0} \dfrac{f(a+h) - f(a)}{h}$ exists but is not finite, or because it does not exist at all.

> **?** The derivative of a function is sometimes represented by $\dfrac{dy}{dx}$ or $\dfrac{d}{dx}(f(x))$. This notation had been introduced by Leibniz long before Cauchy defined derivative in modern terms. While Cauchy's modern definition of derivative is based on the comparison of the variation of two variables at $x = a$, Leibniz looked at it as the ratio of two actual numbers.

Example 11

Show that the function defined by $f(x) = \sqrt{|x|}$ is continuous but not differentiable at $x = 0$.

$\lim_{x \to 0^+} f(x) = \lim_{x \to 0^+} \sqrt{x} = 0$ and

$\lim_{x \to 0^-} f(x) = \lim_{x \to 0^-} \sqrt{-x} = 0.$

$\therefore \lim_{x \to 0} f(x) = 0 = f(0)$

$\lim_{h \to 0^-} \dfrac{f(h) - f(0)}{h} = \lim_{h \to 0^-} \dfrac{\sqrt{-h} - 0}{h} = \lim_{h \to 0^-} \dfrac{1}{\sqrt{-h}} = \infty;$

$\lim_{h \to 0^+} \dfrac{f(h) - f(0)}{h} = \lim_{h \to 0^+} \dfrac{\sqrt{h} - 0}{h} = \lim_{h \to 0^+} \dfrac{1}{\sqrt{h}} = \infty.$

Therefore f is not differentiable at $x = 0$ because the limits above are not finite real numbers.

Use definition of continuous function at $x = 0$

1 $\lim_{x \to 0^-} f(x)$ and $\lim_{x \to 0^+} f(x)$ exist and

$\quad \lim_{x \to 0^-} f(x) = \lim_{x \to 0^+} f(x)$

2 $\lim_{x \to 0} f(x) = f(0)$

Use definition of derivative of f at the point $x = 0$

$\lim_{h \to 0} \dfrac{f(a+h) - f(a)}{h}$

and calculate the left and right limits to check if they exist and are finite, and if so, if they are equal.

Example 12

Show that the function defined by $f(x) = \begin{cases} x\sin\left(\dfrac{1}{x}\right), & x \neq 0 \\ 0, & x = 0 \end{cases}$ is not differentiable at $x = 0$.

$\lim\limits_{h \to 0} \dfrac{f(0+h) - f(0)}{h} = \lim\limits_{h \to 0} \dfrac{h\sin\left(\dfrac{1}{h}\right)}{h} = \lim\limits_{h \to 0} \left(\sin\left(\dfrac{1}{h}\right)\right)$

Use definition of derivative of f at the point $x = 0$.

Let $g(x) = \sin\left(\dfrac{1}{x}\right)$ and consider $u_n = \dfrac{1}{\pi n} \to 0$

and $v_n = \dfrac{2}{\pi(4n+3)} \to 0$, $n \in \mathbb{Z}^+$

As $g(u_n) = \sin(\pi n) = 0$ and $g(v_n) = \sin\left(\dfrac{\pi(4n+3)}{2}\right) = -1$,

so $\lim\limits_{x \to 0} g(x)$ does not exist.

So $\lim\limits_{h \to 0}\left(\sin\left(\dfrac{1}{h}\right)\right)$ does not exist which means that $f'(0)$ does not exist.

*Use result found in example 12 from chapter 1:
find two sequences $\{u_n\}$ and $\{v_n\}$ such that $\lim\limits_{n \to \infty} u_n = \lim\limits_{n \to \infty} v_n = 0$
but $\lim\limits_{n \to +\infty}(g(u_n)) \neq \lim\limits_{n \to +\infty}(g(v_n))$*

As you have studied in the core course, we can establish algebraic properties for derivatives that are summarized by the following theorem:

Theorem 6: Let f and g be differentiable functions at $x = a$. Then:

i $f \pm g$ and fg are differentiable at $x = a$ and

$(f \pm g)'(a) = f'(a) \pm g'(a)$ and

$(fg)'(a) = f'(a)g(a) + f(a)g'(a)$

ii $\dfrac{f}{g}$ is differentiable at $x = a$ and

$\left(\dfrac{f}{g}\right)'(a) = \dfrac{f'(a)g(a) - f(a)g'(a)}{(g(a))^2}$, $g(a) \neq 0$.

iii f^n is differentiable at $x = a$ and

$(f^n)'(a) = n(f(a))^{n-1} f'(a)$, $n \in \mathbb{Z}^+$.

Let f and g be differentiable functions at $g(a)$ and a, respectively.

iv $f \circ g$ differentiable at $x = a$ and $(f \circ g)'(a) = f'(g(a))g'(a)$; if g is the inverse of f, $(f \circ g)'(a) = 1 \Rightarrow g'(a) = \dfrac{1}{f'(g(a))}$.

This is called the chain rule.

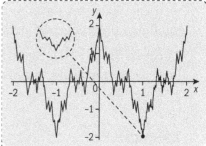

→ At the beginning of the chapter you studied the graph of the Blancmange function that is continuous everywhere but nowhere differentiable. These types of functions are known as pathological functions. Another famous example of pathological behaviour was given by Karl Weierstrass in 1872, when he proposed the function:

$f(x) = \sum\limits_{n=0}^{\infty} \left(\dfrac{2}{3}\right)^n \cos(9^n \pi x)$.

This function expresses f as an infinite, but converging, sum of cosines with ever increasing frequencies. There are an infinite number of 'wiggles', so that it's too 'bumpy' to have a tangent anywhere!

Example 13

Consider the functions defined by $f(x) = \arctan(x)$ for $x \in \mathbb{R}$ and $g(x) = \tan(x)$ for $x \in \left]-\frac{\pi}{2}, \frac{\pi}{2}\right[$. Show that $f'(x) = \dfrac{1}{g'(f(x))}$.

$f'(x) = \dfrac{1}{1+x^2}$ and $g'(x) = \dfrac{1}{\cos^2(x)}$ | Use differentiation rules to find expressions of the derivatives of f and g.

Let $y = f(x)$, i.e. $y = \arctan(x)$ | Change variable taking into account the domain of the variables.

Then, $g'(f(x)) = g'(y) = \dfrac{1}{\cos^2(y)}$ and | Express the derivative of g in terms of y.

$y = \arctan(x) \Rightarrow \tan(y) = x \Rightarrow \cos^2(y) = \dfrac{1}{1+x^2}$ | Use the trigonometric identity

$\therefore g'(f(x)) = 1+x^2$ | $1 + \tan^2(y) = \dfrac{1}{\cos^2(y)}$

Therefore, $f'(x) = \dfrac{1}{1+x^2} = \dfrac{1}{g'(f(x))}$

You have also learnt that derivatives provide an algebraic method to find local maxima and minima of differentiable functions. The following theorem establishes the criteria to determine the maximum and minimum points of a differentiable function.

Theorem 7: Let f be a continuous function on $[a, b]$ and differentiable on $]a, b[$. If the function has a maximum or minimum point at $x = c$ then

i either $f'(c) = 0$ ii or $c = a$ or $c = b$.

> Theorem 7 does not require that the derivative function is defined at the end points of the interval.

Example 14

Consider the function defined by $f(x) = 2\sqrt{2x - x^2}$.
a Show that the largest possible domain of f is $[0, 2]$.
b Show that f is differentiable on $]0, 2[$ but that $f'(x)$ is not defined at the end points of this interval.
c Use a GDC to graph f and state the maximum and minimum points of f on $[0, 2]$.

---

a $D_f = \{x \in \mathbb{R} \mid 2x - x^2 \geq 0\} = [0, 2]$. | Graph $y = 2x - x^2$. The domain is the set of values of x for which the graph is above or on the x-axis.

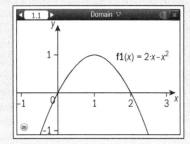

b f is differentiable on $]0, 2[$ because f is the result of the sum, product, and composition of differentiable functions, and $f'(x) = \dfrac{2-2x}{\sqrt{2x-x^2}}$.

f' is not defined at $x = 0$ or at $x = 2$ as these are zeros of the denominator of $f'(x)$.

Use Theorem 6 to justify differentiability of f on $]0, 2[$, find an expression for its derivative, and identify any restrictions on its domain (e.g. denominator cannot be equal to zero)

c

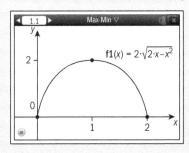

Use a GDC to obtain the graph of f and calculate the maximum point.

The minimum points occur at the end points of the domain.

Maximum at $(1, 2)$; minima at $(0, 0)$ and $(2, 0)$.

Theorems 4 and **7** allow us to prove a very important result: **Rolle's theorem**.

Theorem 8 (Rolle's theorem): Let f be a continuous function on $[a, b]$ and differentiable on $]a, b[$, $a < b$. If $f(a) = f(b)$, then f' has at least one zero on $]a, b[$.

Proof: From Theorem 4, $f([a, b]) = [c, d]$. This means that there is at least one value $x_1 \in [a, b]$ for which $f(x_1) = c$ and one value $x_2 \in [a, b]$ for which $f(x_2) = d$. This means that f has a minimum at x_1 and a maximum at x_2. By Theorem 7, we then have:

i either $f'(x_1) = 0$ or $f'(x_2) = 0$ (or both), for $x_1, x_2 \in]a, b[$ and in this case we have found at least a zero for f' on $]a, b[$ and the theorem is proved;

ii or $x_1, x_2 \in \{a, b\}$. Since $f(a) = f(b)$, f is a constant function and therefore the tangent at any of its points is horizontal and $f'(x) = 0$ for any $x \in]a, b[$.   Q.E.D.

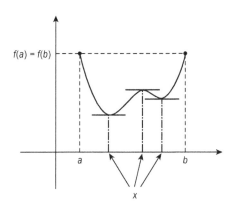

Example 15

Given $f(x) = \cos(2x) + 2\cos(x)$, $0 \leq x \leq 2\pi$, use Rolle's theorem to show that the equation $f'(x) = 0$ has at least one solution on $]0, 2\pi[$. Hence find all the solutions of the equation $f'(x) = 0$ on $]0, 2\pi[$ and verify your answer with your GDC.

---

f is continuous on $[0, 2\pi]$ and differentiable on $]0, 2\pi[$.

Also,
$f(0) = \cos(0) + 2\cos(0) = 1 + 2 = 3$
$f(2\pi) = \cos(2\pi) + 2\cos(2\pi) = 1 + 2 = 3$, so $f(0) = f(2\pi)$.

By Rolle's Theorem, there is at least $x \in]0, 2\pi[$ such that $f'(x) = 0$.

$f'(x) = 0 \Rightarrow -2\sin(2x) - 2\sin(x) = 0$

$-2\sin(x)(2\cos(x) + 1) = 0 \Rightarrow \sin(x) = 0$ or $\cos(x) = -\dfrac{1}{2}$

$\therefore x = \pi$ or $x = \dfrac{2\pi}{3}$ or $x = \dfrac{4\pi}{3}$

Verify conditions of Rolle's Theorem: f is a continuous function on $[a, b]$, differentiable on $]a, b[$, and $f(a) = f(b)$.

Apply conclusion of Rolle's theorem: f' has at least one zero on $]a, b[$.

Use double angle formula and factorize.
Find solution on $]a, b[$.

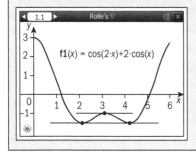

Rolle's theorem is important in calculus because it allows us to prove a fundamental result easily: the Mean Value Theorem.

Theorem 9 (Mean Value Theorem): Let f be a continuous function on $[a, b]$ and differentiable on $]a, b[$. Then, there is at least one value $x \in]a, b[$ such that

$$f'(x) = \frac{f(b) - f(a)}{b - a}.$$

Geometrically, this result tell us that for functions that satisfy the conditions of this theorem, it is always possible to find a tangent to the graph of f between a and b that is parallel to the line AB where A is the point $(a, f(a))$ and B is the point $(b, f(b))$.

Note that in some cases you may be able to find more than one tangent that is parallel to AB, as illustrated in the diagram.

Note that Rolle's Theorem is a particular case of the Mean Value Theorem; when the line AB is horizontal.

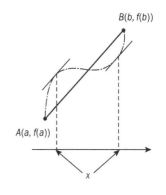

Smoothness in mathematics

Proof of Mean Value Theorem:

Consider the linear function $h(x) = \underbrace{\frac{f(b)-f(a)}{b-a}}_{m}(x-a)$ and the function $g(x) = f(x) - h(x)$.

Theorems 1 and 6 allow us to establish that g is a continuous function on $[a, b]$ and differentiable on $]a, b[$; as g can be seen as the difference between f and a linear function h.

As $g(x) = f(x) - \frac{f(b)-f(a)}{b-a}(x-a)$, we see that $g(a) = g(b) = f(a)$ (you can check it!).

So, we can apply Rolle's Theorem to g: there is at least a value $x \in]a, b[$ such that $g'(x) = 0$.

> Note that the gradient of the line graph of h is the gradient of the line AB shown in the diagram above.

But

$$g'(x) = 0 \Rightarrow f'(x) - \frac{f(b)-f(a)}{b-a}(x-a)' = 0$$

$$\Rightarrow f'(x) - \frac{f(b)-f(a)}{b-a} = 0$$

$$\Rightarrow f'(x) = \frac{f(b)-f(a)}{b-a} \quad \text{Q.E.D}$$

Example 16

Show that $\sqrt{1+h} < 1 + \frac{h}{2}$ for any $h > 0$.	MVT stands for Mean Value Theorem
Given $h > 0$, let $f(x) = \sqrt{1+x}$ for $0 \leq x \leq h$. $f'(x) = \frac{1}{2\sqrt{1+x}}$, for $0 \leq x \leq h$.	*Select an appropriate function and interval where the function satisfies the conditions of the MVT.*
f is continuous on $[0, h]$ and differentiable on $]0, h[$.	*Verify continuity and differentiability of the function.*
Therefore by the MVT, there is at least one value of $x \in]a, b[$ such that $f'(x) = \frac{f(h)-f(0)}{h-0} \Rightarrow \frac{1}{2\sqrt{1+x}} = \frac{\sqrt{1+h}-1}{h}$	*Apply MVT to obtain an equation relating x and h*
As $0 < x < h \Rightarrow 1 + x > 1 \Rightarrow \sqrt{1+x} > 1$ $\Rightarrow \frac{1}{\sqrt{1+x}} < 1 \Rightarrow \frac{1}{2\sqrt{1+x}} < \frac{1}{2}$	*Use the interval chosen to obtain an appropriate inequality and show that the expression of the derivative has an upper bound.*
$\therefore \frac{\sqrt{1+h}-1}{h} < \frac{1}{2} \Rightarrow \sqrt{1+h}-1 < \frac{1}{2}h \Rightarrow \sqrt{1+h} < 1 + \frac{1}{2}h$	*Re-arrange the expression taking into account that $h > 0$.*

The Mean Value Theorem is a powerful result in calculus because it allows us to prove some very important results that you have been using intuitively and that are summarized in the following list of corollaries:

Corollary 1: If $f'(x) = 0$ for all $x \in I \subseteq D_f$, then f is constant on the interval I.

Proof: Consider any two points $a < b$ in I and apply the Mean Value Theorem to f on the interval $[a, b]$. Then

$$0 = f'(x) = \frac{f(b) - f(a)}{b - a} \Rightarrow f(b) = f(a).$$ Thus the function is constant on I.

Corollary 2: If $f'(x) = g'(x)$ for all $x \in I \subseteq D_f \cap D_g$, then $f(x) = g(x) + c$ for all $x \in I$.

The proof of this corollary is left to the student as an exercise.

Corollary 3: If $f'(x) > 0$ for all $x \in I \subseteq D_f$, then f is increasing on the interval I.

Proof: Consider any two points $a < b$ in I. Then

$$f'(x) = \frac{f(b) - f(a)}{b - a} > 0 \Rightarrow f(b) - f(a) > 0 \Rightarrow f(b) > f(a).$$

Therefore, for any two values $b > a \Rightarrow f(b) > f(a)$ which means that f is increasing on I.

Corollary 4: If $f'(x) < 0$ for all $x \in I \subseteq D_f$, then f is decreasing on the interval I.

The proof of this corollary is left to the student as an exercise.

Exercise 2C

1 Let $f(x) = 3|x| + x - 1$, $x \in \mathbb{R}$
   a Show that $f(-2) = f(1)$.
   b Find the derivative function of f and show that f' has no zeros on $]-2, 1[$.
   c Graph f and explain why this function does not meet the requirements of Rolle's theorem on $]-2, 1[$.

2 Consider the polynomial function $f(x) = 2x^5 - 5x^4 - 10x^3 + 10$.
   a Find $f(-1)$ and $f(1)$ and show that f has at least one zero on $]-1, 1[$.
   b Find the derivative function of f and its zeros.
   c Hence show that f cannot have any other zero on $]-1, 1[$.
   d Graph f, showing clearly all its zeros and maxima and minima points.

3 Consider the function $f(x) = x - \ln(x^2 + 3)$

   a Show that the equation $f'(x) = 0$ has no real solution.

   b Hence show that equation $x - \ln(x^2 + 3) = 0$ has at most one real solution.

   c Graph f, showing clearly all its zeros and maxima and minima points.

4 Show that the equation $x^3 + 3x = 2$ has at least one solution between 0 and 1. Use Rolle's theorem to show that this zero is the only real zero.

5 Consider the function $f(x) = \sin(x)$, $x \in \mathbb{R}$.

   a Use the Mean Value Theorem to show $f'(c) = \dfrac{\sin(x)}{x}$ where $c \in \,]0, x[$, for $0 < x < \dfrac{\pi}{2}$.

   b Hence show that $\cos(x) < \dfrac{\sin(x)}{x} < 1$, for $0 < x < \dfrac{\pi}{2}$.

6 Use the Mean Value Theorem applied to suitable functions to show that:

   a $e^x > x + 1$, $x > 0$;

   b $\ln\left(\dfrac{x+1}{x}\right) < \dfrac{1}{x}$, for $x > 0$;

   c $\tan(x) > x$ for $0 < x < \dfrac{\pi}{2}$;

   d $\arcsin(x) > x$, for $0 < x < 1$.

7 Consider the functions $f(x) = \arctan(x)$ and $g(x) = \arctan\left(\dfrac{1+x}{1-x}\right)$, $x > 1$.

Use the Mean Value Theorem to show that $g(x) = f(x) + c$ where c is a real constant.

8 Use the Mean Value Theorem to show that the function defined by $f(x) = x^2 e^{-x}$ is increasing on $]0, 2[$.

> Apply Corollary 1 to a suitable function.

9 Prove Corollary 2 of the Mean Value Theorem.

10 Prove Corollary 4 of the Mean Value Theorem.

> Modify proof of Corollary 3.

Investigation

The Mean Value Theorem provides a method to approximate values of functions: given a differentiable function on $I \in\ ]a, b[$ and $x_1, x_2 \in I$,

$$f(x_2) \approx f(x_1) + (x_2 - x_1) f'(x_1)$$

This approximation of the value of f at x_2 using its value at x_1 is better when the two points are close together. But how accurate is the approximation? Does it depend on the point? Does it depend on the function?

Use a GDC or a spreadsheet to investigate the accuracy of this approximation method, starting with:

a $f(x) = \sqrt{x}$ and $x_1 = 2, 3, 5, 7, 10,...$ and $x_2 = x_1 + h$ with $h = 1, 0.1, 0.01, ...$

b $f(x) = e^x$ and $x_1 = 2, 3, 5, 7, 10,...$ and $x_2 = x_1 + h$ with $h = 1, 0.1, 0.01, ...$

c $f(x) = \ln(x)$ and $x_1 = 2, 3, 5, 7, 10,...$ and $x_2 = x_1 + h$ with $h = 1, 0.1, 0.01, ...$

Write down your conclusions and justify them with reference to the Mean Value Theorem.

2.4 Limits at a point, indeterminate forms, and L'Hopital's rule

In this section we are going to look in detail at three important theorems that follow from Rolle's Theorem. The first one is an extension of the Mean Value Theorem, and its proof shows a clever technique to make a complex proof actually very simple:

Theorem 10: (Cauchy's Theorem): Let f and g be continuous functions on $[a, b]$ and differentiable on $]a, b[$. Then there is at least one value $x \in\ ]a, b[$ such that

$$(f(b) - f(a)) g'(x) = (g(b) - g(a)) f'(x).$$

Proof: Let $h(x) = (f(b) - f(a)) g(x) - (g(b) - g(a)) f(x)$. The function h is clearly continuous on $[a, b]$ and differentiable on $]a, b[$, by theorems 1 and 6. As

$$h(a) = (f(b) - f(a)) g(a) - (g(b) - g(a)) f(a) = f(b)g(a) - f(a)g(b) = h(b),$$

then, by Rolle's theorem there is at least a value $x \in\ ]a, b[$ such that $h'(x) = 0$.

But $h'(x) = 0 \Rightarrow (f(b) - f(a)) g'(x) - (g(b) - g(a)) f'(x) = 0$
$\Rightarrow (f(b) - f(a)) g'(x) = (g(b) - g(a)) f'(x).$ Q.E.D.

> As the Mean Value Theorem was used to prove Cauchy's Theorem, we cannot use Cauchy's Theorem to prove the Mean Value Theorem as this would be circular reasoning, and therefore not valid.

The Mean Value Theorem is a special case of Cauchy's Theorem.
When we take $g(x) = x$ in Cauchy's Theorem, we see that:

$g(x) = x \Rightarrow g'(x) = 1$ and therefore we obtain
$(f(b) - f(a))g'(x) = (g(b) - g(a))f'(x) \Rightarrow (f(b) - f(a)) \cdot 1 = (b - a)f'(x)$

and thus $f'(x) = \dfrac{f(b) - f(a)}{b - a}$. This is the Mean Value Theorem.

The importance of Cauchy's Theorem is highlighted by the following result we can derive from it:

$(f(b) - f(a))g'(x) = (g(b) - g(a))f'(x) \Rightarrow \dfrac{f'(x)}{g'(x)} = \dfrac{f(b) - f(a)}{g(b) - g(a)}$

for $g(a) \neq g(b)$ and $g'(x) \neq 0$.

Suppose that $\lim_{x \to a} f(x) = \lim_{x \to a} g(x) = 0$. Then $\lim_{x \to a} \dfrac{f(x)}{g(x)} = \dfrac{0}{0}$ which is an indeterminate form.

But, if we apply Cauchy's Theorem to the intervals $]a - h, a[$ and/or $]a, a + h[$ it can be shown that $\lim_{x \to a} \dfrac{f(x)}{g(x)} = \lim_{x \to a} \dfrac{f'(x)}{g'(x)}$.

> We can just use the L'Hôpital rule when there is an appropriate indeterminate form.

The details of this proof are beyond the scope of the Higher Level course, but the conditions and result are given by the following theorem:

L'Hôpital Rule: If $\lim_{x \to a} f(x) = \lim_{x \to a} g(x) = 0$ and $\lim_{x \to a} \dfrac{f'(x)}{g'(x)}$ exists, then $\lim_{x \to a} \dfrac{f(x)}{g(x)}$ also exists, and $\lim_{x \to a} \dfrac{f(x)}{g(x)} = \lim_{x \to a} \dfrac{f'(x)}{g'(x)}$.

Example 17

Apply L'Hôpital Rule to calculate the following limit $\lim_{x \to 0} \dfrac{e^{2x} - e^{-x}}{\sin x}$	
$\lim_{x \to 0} \dfrac{e^{2x} - e^{-x}}{\sin x} = \dfrac{1-1}{0} = \dfrac{0}{0}$ which is an indeterminate form	Let $f(x) = e^{2x} - e^{-x}$ and $g(x) = \sin x$
As $\dfrac{d}{dx}(e^{2x} - e^{-x}) = 2e^{2x} + e^{-x}$ and $\dfrac{d}{dx}(\sin x) = \cos x$ $\lim_{x \to 0} \dfrac{2e^{2x} + e^{-x}}{\cos x} = \dfrac{2+1}{1} = 3$, and by L'Hôpital Rule: $\lim_{x \to 0} \dfrac{e^{2x} - e^{-x}}{\sin x} = 3$	Verify that $\lim_{x \to 0} \dfrac{f'(x)}{g'(x)}$ exists before applying L'Hôpital Rule.

We can only apply L'Hôpital Rule when you have an indeterminate form.
The example below shows that the result is not valid for other situations:

Example 18

> Consider the functions defined by $f(x) = x + \cos(x)$ and $g(x) = x$.
> Show that $\lim\limits_{x \to 0} \dfrac{f(x)}{g(x)} \neq \lim\limits_{x \to 0} \dfrac{f'(x)}{g'(x)}$ and explain why L'Hôpital Rule cannot be applied to calculate $\lim\limits_{x \to 0} \dfrac{f(x)}{g(x)}$.

$\lim\limits_{x \to 0} \dfrac{f(x)}{g(x)} = \lim\limits_{x \to 0} \dfrac{x + \cos(x)}{x} = \dfrac{1}{0} = \infty$	Use algebra of limits rules.
$\lim\limits_{x \to 0} \dfrac{f'(x)}{g'(x)} = \lim\limits_{x \to 0} \dfrac{1 - \sin(x)}{1} = 1$	
L'Hôpital Rule cannot be applied to calculate $\lim\limits_{x \to 0} \dfrac{f(x)}{g(x)}$ because $\lim\limits_{x \to 0} f(x) \neq 0$.	Read carefully the conditions of L'Hôpital Rule.

L'Hôpital Rule can be **extended** to include limits that can be reduced to the form $\lim\limits_{x \to \infty} \dfrac{f(x)}{g(x)}$:

> The proof of this result is beyond the scope of the course.

If $\lim\limits_{x \to \infty} \dfrac{f(x)}{g(x)} = \dfrac{0}{0}$ or $\lim\limits_{x \to \infty} \dfrac{f(x)}{g(x)} = \dfrac{\infty}{\infty}$ and $\lim\limits_{x \to \infty} \dfrac{f'(x)}{g'(x)}$ exists, then $\lim\limits_{x \to \infty} \dfrac{f(x)}{g(x)}$ also exists and $\lim\limits_{x \to \infty} \dfrac{f(x)}{g(x)} = \lim\limits_{x \to \infty} \dfrac{f'(x)}{g'(x)}$.

The following example shows you how to apply this extended version of L'Hôpital Rule.

Example 19

> Apply L'Hôpital Rule to calculate the following limit
> **a** $\lim\limits_{x \to \infty} \dfrac{e^{2x}}{x}$ **b** $\lim\limits_{x \to \infty} \dfrac{\ln x}{\sqrt{x}}$

a $\lim\limits_{x \to +\infty} \dfrac{e^{2x}}{x} = \dfrac{\infty}{\infty}$ indeterminate form As $\lim\limits_{x \to +\infty} \dfrac{2e^{2x}}{1} = \dfrac{+\infty}{1} = +\infty$, by L'Hôpital Rule, $\lim\limits_{x \to +\infty} \dfrac{e^{2x}}{x} = +\infty$	Let $f(x) = e^{2x}$ and $g(x) = x$ and verify that $\lim\limits_{x \to 0} \dfrac{f'(x)}{g'(x)}$ exists before applying L'Hôpital Rule.
b $\lim\limits_{x \to +\infty} \dfrac{\ln x}{\sqrt{x}} = \dfrac{\infty}{\infty}$ As $\lim\limits_{x \to +\infty} \dfrac{\frac{1}{x}}{\frac{1}{2\sqrt{x}}} = \lim\limits_{x \to \infty} \dfrac{2}{\sqrt{x}} = 0$, by L'Hôpital Rule, $\lim\limits_{x \to +\infty} \dfrac{\ln x}{\sqrt{x}} = 0$	Let $f(x) = \ln x$ and $g(x) = \sqrt{x}$ and verify that $\lim\limits_{x \to 0} \dfrac{f'(x)}{g'(x)}$ exists before applying L'Hôpital Rule.

Exercise 2D

1 Use L'Hôpital Rule to find the following limits:

a $\lim\limits_{x \to 0} \dfrac{x^2 - 3x}{\sin x}$

b $\lim\limits_{x \to 0} \dfrac{e^{2x} - e^{-3x}}{x}$

c $\lim\limits_{x \to 0} \dfrac{\cos(x) + 2x - 1}{3x}$

d $\lim\limits_{x \to 0} \dfrac{e^{2x} + e^{-x} - 2}{1 - \cos(2x)}$

e $\lim\limits_{x \to \frac{\pi}{2}^+} \dfrac{\ln(\cot x)}{\ln(\cos x)}$

f $\lim\limits_{x \to 0} \dfrac{x(1 - e^x)}{\ln(1 - x)}$

i $\lim\limits_{x \to \pi} \dfrac{1 + \cos(x)}{x^2 \sin(x)}$

g $\lim\limits_{x \to 0} \dfrac{x - \arcsin(x)}{\sin^2(x)}$

h $\lim\limits_{x \to 0} \dfrac{e^{2x} - \cos(3x)}{x \sin(2x)}$

2 The L'Hôpital Rule was used to find the limits below. Explain why the method used is not a valid method. Show that the values found are incorrect and state the correct value for each limit.

a $\lim\limits_{x \to 1} \dfrac{x^3 + 3x - 2}{x^2 - x + 2} = \lim\limits_{x \to 1} \dfrac{3x^2 + 3}{2x - 1} = \dfrac{6}{1} = 6$

b $\lim\limits_{x \to 1} \dfrac{1 - \cos(x)}{x^2 + x} = \lim\limits_{x \to 1} \dfrac{\sin(x)}{2x + 1} = \lim\limits_{x \to 1} \dfrac{\cos(x)}{2} = \dfrac{\cos(1)}{2}$

3 Use L'Hôpital Rule to find the following limits:

a $\lim\limits_{x \to +\infty} \dfrac{e^{5x} + x}{4x + 1}$

b $\lim\limits_{x \to +\infty} \dfrac{x^2}{x(e^x + 1)}$

c $\lim\limits_{x \to +\infty} \dfrac{\ln(e^x + 1)}{x + 3}$

EXTENSION QUESTIONS

4 Consider the functions defined by $f(x) = \dfrac{1}{x}$ and $g(x) = (x - 1)^3$.

a Show that f and g satisfy the conditions of Cauchy's Theorem.

b Hence, show that equation $(f(3) - f(1))g'(x) = (g(3) - g(1))f'(x)$ has at least one solution on the interval $]1, 3[$. Solve the equation and determine the number of solutions on $]1, 3[$.

5 Consider the functions defined by $f(x) = x^3 - 3x$ and $g(x) = (x - 1)^2$.

a Show that f and g satisfy the conditions of the Mean Value Theorem.

b Use Cauchy's Theorem to show that
$(f(2) - f(-4))g'(x) = (g(2) - g(-4))f'(x)$ has a solution on $]-4, 2[$.

c Show that equation $\dfrac{g'(x)}{f'(x)} = \dfrac{g(2) - g(-4)}{f(2) - f(-4)}$ has no solution on the interval $]-4, 2[$.

d Comment on your answers to parts **b** and **c**.

Other indeterminate forms

Using algebraic manipulations similar to the ones you explored in chapter 1, you can transform expressions of functions and convert the following indeterminate forms into $\frac{0}{0}$ form:

$\frac{\infty}{\infty}$, $\infty - \infty$, $0 \times \infty$, 0^0, ∞^0, and 1^∞.

The following examples show how to deal with some of these types of indeterminate forms. In some cases you may need to apply L'Hôpital Rule more than once to determine the value of the limit.

Example 20

Calculate $\lim_{x \to 1} \left(\dfrac{x}{x-1} - \dfrac{1}{\ln x} \right)$.

$\lim_{x \to 1} \left(\dfrac{x}{x-1} - \dfrac{1}{\ln x} \right) = \pm\infty - (\pm\infty)$	Identify the indeterminate form
$\lim_{x \to 1} \left(\dfrac{x}{x-1} - \dfrac{1}{\ln x} \right) = \lim_{x \to 1} \left(\dfrac{x \ln x - (x-1)}{(x-1)\ln x} \right) = \dfrac{0}{0}$	Convert it into $\dfrac{0}{0}$
$\lim_{x \to 1} \left(\dfrac{\ln x + x \cdot \frac{1}{x} - 1}{\ln x + (x-1) \cdot \frac{1}{x}} \right) = \dfrac{0}{0}$	Apply L'Hôpital Rule, check conditions of new functions and apply L'Hôpital Rule again.
$\lim_{x \to 1} \left(\dfrac{\ln x + x \cdot \frac{1}{x} - 1}{\ln x + (x-1) \cdot \frac{1}{x}} \right) = \lim_{x \to 1} \left(\dfrac{\ln x}{\ln x + (x-1) \cdot \frac{1}{x}} \right) = \dfrac{0}{0}$	
$= \lim_{x \to 1} \left(\dfrac{\frac{1}{x}}{\frac{1}{x} + 1 \cdot \frac{1}{x} + (x-1) \cdot \left(-\frac{1}{x^2}\right)} \right) = \lim_{x \to 1} \dfrac{1}{1+1+0} = \dfrac{1}{2}$	

Example 21

Calculate $\lim\limits_{x\to\frac{\pi}{2}}\left(\cos(x)\cdot\dfrac{1}{\ln\left(\frac{2}{\pi}x\right)}\right)$.	
$\lim\limits_{x\to\frac{\pi}{2}}\left(\cos(x)\cdot\dfrac{1}{\ln\left(\frac{2}{\pi}x\right)}\right) = 0\times\infty$	Identify the indeterminate form
$\lim\limits_{x\to\frac{\pi}{2}}\left(\dfrac{\cos(x)}{\ln\left(\frac{2}{\pi}x\right)}\right) = \dfrac{0}{0}$	Convert it into $\dfrac{0}{0}$
$\lim\limits_{x\to\frac{\pi}{2}}\left(\dfrac{\cos(x)}{\ln\left(\frac{2}{\pi}x\right)}\right) = \lim\limits_{x\to\frac{\pi}{2}}\left(\dfrac{-\sin(x)}{\frac{2}{\pi}\cdot\frac{1}{\frac{2}{\pi}x}}\right)$ $= \lim\limits_{x\to\frac{\pi}{2}}(-x\sin(x)) = -\dfrac{\pi}{2}$	Check conditions of new functions and apply L'Hôpital Rule to the new limit.

Example 22

Calculate $\lim\limits_{x\to 0^+}\left((x+e^{2x})^{\frac{1}{x}}\right)$.	
$\lim\limits_{x\to 0^+}\left((x+e^{2x})^{\frac{1}{x}}\right) = 1^\infty$	Identify the indeterminate form
$\lim\limits_{x\to 0^+}\left((x+e^{2x})^{\frac{1}{x}}\right) = e^{\lim\limits_{x\to 0^+}\left(\frac{\ln(x+e^{2x})}{x}\right)}$	Use the fact that for continuous functions f and g $\lim\limits_{x\to a}(f(x)^{g(x)}) = e^{\lim\limits_{x\to a}(g(x)\cdot\ln(f(x)))}$
Now $\lim\limits_{x\to 0^+}\left(\dfrac{\ln(x+e^{2x})}{x}\right) = \dfrac{0}{0}$ and $\lim\limits_{x\to 0^+}\left(\dfrac{\ln(x+e^{2x})}{x}\right) = \lim\limits_{x\to 0^+}\left(\dfrac{1+2e^{2x}}{x+e^{2x}}\right) = 3$ $\therefore \lim\limits_{x\to 0^+}\left((x+e^{2x})^{\frac{1}{x}}\right) = e^3$	Check conditions of new functions and apply L'Hôpital Rule.

Exercise 2E

1 Identify the indeterminate form, transform it into $\frac{0}{0}$ or $\frac{\infty}{\infty}$, and use L'Hôpital Rule to find the following limits:

a $\lim\limits_{x \to 0^+} \left(\frac{3}{x} e^{-\frac{2}{x}} \right)$

b $\lim\limits_{x \to 1^+} \left((\arctan(x-1)) e^{\frac{3}{x-1}} \right)$

c $\lim\limits_{x \to 0^+} \left(\sin(x) \cdot (\ln(x) - \cot(x)) \right)$

d $\lim\limits_{x \to 0^+} (x + e^x)^{\frac{2}{x}}$

> Note that you may need to apply L'Hôpital Rule more than once to calculate the limits given.

e $\lim\limits_{x \to 0^+} (\sin(2x))^{3x}$

f $\lim\limits_{x \to 0} (1 + x)^{\frac{5}{x}}$

g $\lim\limits_{x \to 0} (2x \cot(5x))$

h $\lim\limits_{x \to 0} (\ln(e - x))^{\cot(x)}$

i $\lim\limits_{x \to 0} (e^{e^x - 1})^{\frac{1}{x}}$

k $\lim\limits_{x \to 0^+} (\ln(x) - \ln(\sin(x)))$

l $\lim\limits_{x \to -\infty} x e^x$

m $\lim\limits_{x \to +\infty} (\ln(e^x + 1) - \ln(e^x - 1))$

n $\lim\limits_{x \to +\infty} \left(x \sin\left(\frac{1}{x}\right) \right)$

o $\lim\limits_{x \to +\infty} \left((1 + x)^{\frac{1}{x}} \right)$

2 Consider the piecewise functions defined by

a $f(x) = \begin{cases} \frac{\sin(3x)}{x}, & x \neq 0 \\ a, & x = 0 \end{cases}$

b $g(x) = \begin{cases} (e^x + 4x)^{\frac{1}{x}}, & x > 0 \\ b - x, & x \leq 0 \end{cases}$

c $h(x) = \begin{cases} \frac{1 - \cos(x)}{x^2}, & x \neq 0 \\ c, & x = 0 \end{cases}$

Find the values of a, b and c that make each of the functions continuous.

3 Consider the function $f(x) = \left(1 + \frac{1}{x}\right)^x$, $x > 0$.

a Graph f and state the value the graph suggests for $\lim\limits_{x \to +\infty} f(x)$

b Consider the function $g(x) = \ln(f(x))$. Use the substitution $h = \frac{1}{x}$ and express g as a function of h.

c Hence calculate $\lim\limits_{h \to 0^+} g(h)$ and use it to find $\lim\limits_{x \to +\infty} f(x)$.
Compare your result with your answer to part **a**.

Smoothness in mathematics

2.5 What are *smooth graphs* of functions?

In this chapter we have explored examples of functions that are continuous but not differentiable at a point. However, an important consequence of L'Hôpital Rule is a result about the relation between continuity of a function, and continuity of its derivative:

Let f be a continuous function for $x = a$ and suppose that f' exists in an interval that contains a, but f' does not exist at a.
Then $f'(a) = \lim_{x \to a} \dfrac{f(x)-f(a)}{x-a} = \lim_{x \to a} \dfrac{f'(x)}{1} = \lim_{x \to a} f'(x)$ when this limit exists. This is a contradiction, as we had assumed that $f'(a)$ did not exist.

Thus, the graph of the derivative of a continuous function cannot have a 'hole'. Either the derivative f' of the function f exists everywhere on the interval, or else the function f is not continuous.

In mathematics we say the graph of a function is **smooth** when both the function and its derivative are continuous functions.

For example, $f(x) = \begin{cases} x^2 \sin\left(\dfrac{1}{x}\right), & x \neq 0 \\ 0, & x = 0 \end{cases}$ is continuous at $x = 0$ but

$\lim_{x \to 0} f'(x) = \lim_{x \to 0} \left(2x \sin\left(\dfrac{1}{x}\right) - \sin\left(\dfrac{1}{x}\right) \right)$ does not exist and therefore the graph of f is not smooth.

In general, we just need to study smoothness of graphs of functions (or curves) for piecewise functions at the end points of each branch. We can also determine values for parameters to make the graph of a piecewise function smooth; as shown in the example below.

Example 23

Let f be a function defined by $f(x) = \begin{cases} x+a, & x \geq 1 \\ x^2 + bx, & x < 1 \end{cases}$

Find the value of a and b such that the function f is continuous and its derivative exists and is also continuous.

$\lim_{x \to 1^-} f(x) = 1 + a$ and $\lim_{x \to 1^+} f(x) = 1 + b$ $\therefore a = b$	f is continuous at $x = 1$ only if $\lim_{x \to 1^-} f(x) = \lim_{x \to 1^+} f(x)$
$f'(x) = \begin{cases} 1, & x > 1 \\ 2x + b, & x < 1 \end{cases}$ f' exists at $x = 1$ when $\lim_{x \to 1^+} (2 + b) = 1 \Rightarrow b = -1$ $\therefore a = b = -1$	Determine expressions for the derivative of f at all points except $x = 1$ use consequence of L'Hôpital Rule to establish the existence of derivative at $x = 1$.

Exercise 2F

For each of the following piecewise functions find, if possible, the values of a and b such that the function is continuous and its derivative exists and is also continuous.

1 $f(x) = \begin{cases} ax^2 - x, & x > 2 \\ x^3 + bx, & x \leq 2 \end{cases}$

2 $f(x) = \begin{cases} \dfrac{e^{ax} - 1}{x}, & x > 0 \\ 3x + b, & x \leq 0 \end{cases}$

3 $f(x) = \begin{cases} \ln(ax), & x > 1 \\ \arctan(bx), & x \leq 1 \end{cases}$

2.6 Limits of functions and limits of sequences

In chapter 1, we learnt how to relate limits of sequences to limits of functions. In particular, we saw that $\lim_{x \to +\infty} f(x)$ exists and $\lim_{x \to +\infty} f(x) = L$ if and only if for any sequence $\{a_n\}$ such that $a_n \in I$ for all $n \in \mathbb{Z}^+$ and $\lim_{n \to \infty} a_n = +\infty$, $\lim_{n \to \infty}(f(a_n)) = L$.

However, if we know how to find the value of $\lim_{x \to +\infty} f(x)$ we can use it to find the limit of any sequence such that $u_n = f(n)$ for $n \geq N$ where $n \in \mathbb{Z}^+$ (i.e. the sequence is a restriction of f to natural numbers greater than or equal to N).

This allows us to apply L'Hôpital Rule to find limits of sequences when indeterminate forms occur.

Example 24

Find, if possible, $\lim_{n \to \infty} \dfrac{3n^2 + n}{e^n}$.

$\lim_{n \to \infty} \dfrac{3n^2 + n}{e^n} = \dfrac{\infty}{\infty}$ is an indeterminate form.

Consider $f(x) = 3x^2 + x$ and $g(x) = e^x$.

$\lim_{x \to +\infty} \dfrac{f(x)}{g(x)} = \underbrace{\lim_{x \to +\infty} \dfrac{6x + 1}{e^x}}_{\lim_{x \to \infty} \frac{f'(x)}{g'(x)}} = \underbrace{\lim_{x \to +\infty} \dfrac{6}{e^x}}_{\lim_{x \to \infty} \frac{f''(x)}{g''(x)}} = \dfrac{6}{+\infty} = 0$

$\therefore \lim_{n \to \infty} \dfrac{3n^2 + n}{e^n} = 0$

Verify that you have an indeterminate form. Select appropriate functions to apply L'Hôpital Rule (twice)

The limit of the sequence exists and is equal to $\lim_{x \to +\infty} \dfrac{f(x)}{g(x)}$ as the sequence is given by $\left(\dfrac{f}{g}\right)(n)$.

> Note that the limit in this example cannot be found using the methods studied in chapter 2. However at the end of chapter 5 you will learn an alternative method to deal with indeterminate forms as L'Hôpital Rule is not always the adequate or most efficient method to calculate limits.

Exercise 2G

Find, if possible, the following limits:

1 $\lim_{n \to +\infty} \dfrac{n^2 + 2n + 1}{e^{3n} + n}$

2 $\lim_{n \to +\infty} \dfrac{\arctan\left(\dfrac{1}{n}\right)}{\ln\left(1 + \dfrac{2}{n}\right)}$

3 $\lim_{n \to +\infty} \left(\dfrac{n}{2^n}\right)^{\frac{1}{n}}$

Review exercise

EXAM-STYLE QUESTIONS

1. Consider the equation $2 \arctan x = x$.
   a. Use Rolle's Theorem to show that this equation has at most three real solutions.
   b. Hence show that this equation has exactly one solution on $]-1, 1[$.

2. Consider the function defined by $f(x) = \cos(x) - 2x$, $x \in \mathbb{R}$.
   a. Show that f has exactly one zero on $\left]0, \dfrac{\pi}{4}\right[$, stating clearly any theorem used.
   b. Find $\lim\limits_{x \to 0} \left(f(x)^{\frac{1}{x}}\right)$.
   c. State $\lim\limits_{x \to +\infty} f(x)$.

3. Use the Mean Value Theorem to show that
   a. $|\sin x - \sin y| \leq |x - y|$
   b. $\sqrt{x} - \dfrac{x-y}{2\sqrt{y}} < \sqrt{y} < \sqrt{x} - \dfrac{x-y}{2\sqrt{x}}$, $y > x > 0$.
   c. $\ln\left(\dfrac{x}{x-1}\right) < \dfrac{1}{x-1}$, $x > 1$
   d. $\sin x \leq x$, $0 \leq x \leq \dfrac{\pi}{2}$.

4. Show that $\lim\limits_{x \to 1} \dfrac{x^2 - x}{\cot\left(\dfrac{\pi x}{2}\right)} = -\dfrac{2}{\pi}$

5. Given the function defined by $f(x) = \cot\left(\dfrac{\pi x}{2}\right)$, $x \in]0, \pi[$ calculate
   a. $\lim\limits_{x \to 0} (x \cdot f(x))$
   b. $\lim\limits_{x \to 0} \left(\ln(e + x)^{f(x)}\right)$

6. Let f be a differentiable function defined on $[-a, a]$, $a \in \mathbb{R}^+$.
   a. Use the definition of derivative of a function to show that:
      i. if f is an even function then f' is an odd function;
      ii. if f is an odd function then f' is an even function.
   b. Use the Mean Value Theorem to show that:
      i. if f is an even function then f' has at least one maximum or minimum point on the interval $]-a, a[$;
      ii. if f is an odd function then f has at least one point of inflexion on $]-a, a[$.
   c. Show that if f is odd its graph meets the line $y = x$
      i. at least once, and state the coordinates of this point;
      ii. an odd number of times in the interval $[-a, a]$
   d. Sketch a graph of an odd function defined on the interval $[-a, a]$ such that its graph meets the line $y = x$ exactly three times.
   e. Hence show that the tangents to at least two points on the graph are parallel to the line $y = x$.

Chapter 2 summary

Definition: Given $a \in D_f$, if $\lim_{x \to a} f(x) = f(a)$ then f is continuous at the point a. If f is continuous at each point on an interval I, we say that the functions is continuous on I. A functions is discontinuous at a point $x = a$ when:

1. $\lim_{x \to a^-} f(x) \neq \lim_{x \to 1^+} f(x)$, and in this case we say that $\lim_{x \to a} f(x)$ does not exist;

2. $\lim_{x \to a^-} f(x) = \lim_{x \to a^+} f(x)$ (i.e., $\lim_{x \to a} f(x)$ exists) but $\lim_{x \to a} f(x) \neq f(a)$

Theorems about continuous functions:

T1: Let f and g be continuous functions on their domains. Then:

i $f \pm g$ and fg are continuous on $D = D_f \cap D_g$;

ii $\dfrac{f}{g}$ is continuous on $D = (D_f \cap D_g) \setminus \{x \mid g(x) \neq 0\}$;

iii f^n is continuous on D_f;

iv $\sqrt[n]{f}$ is continuous on D_f when n is odd, and on $D_f \cap \{x \mid f(x) \geq 0\}$ when n is even;

v $f \circ g$ is continuous on $D = \{x \in D_g \mid g(x) \in D_f\}$.

T2: Let f be a continuous one-one function on its domain. Then f^{-1} is also continuous on its domain.

T3: Let f be a continuous one-one function on $I \subset D_f$. Then both f and f^{-1} are either increasing or decreasing on I.

Definition: A continuous function f is **differentiable** at $x = a$ if and only if the $\lim_{h \to 0} \dfrac{f(x+h) - f(a)}{h}$ exists and is finite. In this case we represent this limit by $f'(a)$ and call it the derivative of f at $x = a$. If a function is differentiable at all points on an interval I we say that the function is differentiable on I and we define the derivative function of f as $f'(x) = \lim_{h \to 0} \dfrac{f(x+h) - f(x)}{h}, x \in I.$

Main theorems about differentiable functions

Theorem: Let f and g be differentiable functions at $x = a$. Then:

i $f \pm g$ and fg are differentiable at $x = a$ and $(f \pm g)'(a) = f'(a) \pm g'(a)$ and
    $(fg)'(a) = f'(a)g(a) - f(a)g'(a)$

ii $\dfrac{f}{g}$ is differentiable at $x = a$ and $\left(\dfrac{f}{g}\right)'(a) = \dfrac{f'(a)g(a) - f(a)g'(a)}{(g(a))^2}$, $g(a) \neq 0$.

iii f^n is differentiable at $x = a$ and $(f^n)'(a) = n(f(a))^{n-1} f'(a)$.

Chain Rule: Let f and g be differentiable functions at $g(a)$ and a, respectively. $f \circ g$ differentiable at $x = a$ and $(f \circ g)'(a) = f'(g(a))g'(a)$

Derivative of inverse function: if g is the inverse of f, then
$(f \circ g)'(a) = 1 \Rightarrow g'(a) = \dfrac{1}{f'(g(a))}$.

Criteria for maxima/minima: Let f be a continuous function on $[a, b]$ and differentiable on $]a, b[$. If the function has a maximum or minimum point at $x = c$ then either $f'(c) = 0$ or $c \in \{a, b\}$.

Weierstrass' Theorem: Let f be continuous on $[a, b] \subset D_f$, $a < b$. Then $f([a, b]) = [c, d]$, $c < d$.

Bolzano's Theorem: Let f be continuous on $[a, b] \subset D_f$. If $f(a) < c < f(b)$ or $f(b) < c < f(a)$, then the equation $f(x) = c$ has at least one solution on $[a, b]$.

Corollary: Let f and g be a continuous on $[a, b]$. If $f(a) < g(a)$ and $g(b) < f(b)$, then the equation $f(x) = g(x)$ has at least one solution on $[a, b]$.

Rolle's Theorem: Let f be a continuous function on $[a, b]$ and differentiable on $]a, b[$, $a < b$. If $f(a) = f(b)$, then f' has at least one zero on $]a, b[$.

Mean Value Theorem: Let f be a continuous function on $[a, b]$ and differentiable on $]a, b[$. Then, there is at least one value $x \in]a, b[$ such that $f'(x) = \dfrac{f(b) - f(a)}{b - a}$.

Corollary 1: If $f'(x) = 0$ for all $x \in I \subseteq D_f$, then f is constant on the interval I.

Corollary 2: If $f'(x) = g'(x)$ for all $x \in I \subseteq D_f \cap D_g$, then $f(x) = g(x) + c$ for all $x \in I$.

Corollary 3: If $f'(x) > 0$ for all $x \in I \subseteq D_f$, then f is increasing on the interval I.

Cauchy's theorem: Let f and g be continuous functions on $[a, b]$ and differentiable on $]a, b[$. Then, there is at least a value $x \in]a, b[$ such that
$(f(b) - f(a))g'(x) = (g(b) - g(a))f'(x)$.

L'Hôpital rule:

If $\lim\limits_{x \to a} f(x) = \lim\limits_{x \to a} g(x) = 0$ and $\lim\limits_{x \to a} \dfrac{f'(x)}{g'(x)}$ exists, then $\lim\limits_{x \to a} \dfrac{f(x)}{g(x)}$ also exists, and $\lim\limits_{x \to a} \dfrac{f(x)}{g(x)} = \lim\limits_{x \to a} \dfrac{f'(x)}{g'(x)}$.

Algebraic manipulations that allow application of L'Hôpital Rule to other indeterminate forms:

- Transform the forms $\dfrac{\infty}{\infty}$, $\infty - \infty$, $0 \times \infty$, 0^0, ∞^0 and 1^∞ into the indeterminate form $\dfrac{0}{0}$.

- When the functions f and g are continuous use $\lim\limits_{x \to a}\left(f(x)^{g(x)}\right) = e^{\lim\limits_{x \to a}(g(x) \cdot \ln(f(x)))}$

Modeling dynamic phenomena

CHAPTER OBJECTIVES:

9.5 First order differential equations. Geometric interpretation using slope fields, including identification of isoclines. Numerical solution of $\frac{dy}{dx} = f(x, y)$ using Euler's method. Variables separable differential equations. Homogeneous differential equation $\frac{dy}{dx} = f\left(\frac{y}{x}\right)$ using the substitution $y = vx$. Solution of $y' + P(x)y = Q(x)$, using the integrating factor.

Before you start

You should know how to:

1. Solve equations for a given variable.
   e.g. Solve for y: $xe^{2y} - (x - \sin x) = x^2$.
   $xe^{2y} = x^2 + x - \sin x \Rightarrow e^{2y} = \frac{x^2 + x - \sin x}{x}$
   $\Rightarrow y = \frac{1}{2}\ln\left(\frac{x^2 + x - \sin x}{x}\right)$

2. Find integrals and derivatives of polynomial, rational, irrational, exponential, logarithmic, and trigonometric functions.
   e.g. Differentiate $y = e^x + \sin x + \frac{x}{x^2 - 1}$.
   Using addition and product rules,
   $\frac{dy}{dx} = e^x + \cos x + \frac{1 \cdot (x^2 - 1) - x \cdot 2x}{(x^2 - 1)^2}$
   $= e^x + \cos x - \frac{x^2 + 1}{(x^2 - 1)^2}$

3. Find and interpret the gradient of lines.
   e.g. The gradient of a curve is given by $\frac{dy}{dx} = (x+3)e^x$, determine its exact value when $x = 2$.
   Substitute $x = 2$ into the equation
   $\frac{dy}{dx} = ((2) + 3)e^2 = 5e^2$

Skills check:

1. Solve for y:
   a $xe^y + x^2 = 1$ b $\ln\left(\frac{x+1}{y}\right) = 2x$
   c $\tan\left(\frac{x}{y}\right) = 1$ d $\arccos(xy - 1) = 2x$

2. a Find the following integrals:
   i $\int x^2 e^x dx$ ii $\int \sin(2x)\, e^x dx$
   iii $\int \frac{e^x}{e^x - 1} dx$ iv $\int \frac{\cos(x)}{\sin^2(x)} dx$

   b Find the derivatives of these functions:
   i $y = \arcsin\left(\frac{x}{x-1}\right)$ ii $y = \ln(\sin^2(x) + 1)$

3. If the gradient of a line is given by $\frac{dy}{dx}$, determine its exact value when:
   a $\frac{dy}{dx} = e^{x+1} - \ln(x)$ and $x = 1$
   b $\frac{dy}{dx} = \sin(2x)$ and $x = \frac{\pi}{3}$

Introduction to differential equations

November 7th, 1940 is a well-known date for many engineers. On this day 'Gertie', the galloping bridge in the state of Washington in the USA, collapsed due to the action of wind. Many years later another strange phenomenon occured. The Millennium Bridge in London, a 320 m lateral suspension bridge connecting London's financial district to Bankside, had just been inaugurated. Thousands of pedestrians streamed over it. At first the bridge was still, but then it began to sway. Initially it only swayed just slightly. Then, almost from one moment to the next, the wobble intensified, and suddenly people were planting their feet wide; trying to balance themselves, and pushing out to the side in unison. The effect was dramatic and the bridge was closed almost immediately.

What do these cases have in common? Why did the models used by the designers fail?

In this chapter we are going to explore new modeling techniques that take into account variations of the modeling function. As the examples of the two bridges described above show, sometimes these variations cannot be neglected. However, it is very difficult for scientists and engineers to consider all the factors that may affect the model without making its study too complex and its mathematical analysis extremely difficult. For this reason, in real-life scientists are often forced to compromise between a model that may be easy to

> ❓ Calculus is the branch of mathematics that defines and deals with limits, derivatives and integrals of functions. Traditionally it was divided into two parts: *Differential* and *Integral*, however other branches which use methods from both, such as *Differential Equations*, grew into separate disciplines within calculus.

analyze but may not take into account all the factors affecting the phenomenon in study, and a more complex model that may suit the situation better but is difficult to analyze. Modeling dynamic phenomena is still a challenge for mathematicians nowadays despite the developments in technology that make simulations possible.

To clarify this topic, let's look at the family of equations that mathematicians use to describe dynamic phenomena. They are called *differential equations*. But what *are* differential equations? They are complex equations where the unknown is a function and the equation describes a relation between: this function, its derivatives, the independent variables, and some constants. Even in simple cases where the function depends on a single independent variable, such as $y = f(x)$, the model may look complicated. For example, the equation that models the shape of a cable hanging with its ends fastened is given by

$$\frac{y''}{\sqrt{1+(y')^2}} = k$$

where the constant k depends on the size of the cable. This is in fact a differential equation that mathematicians have no problems with. The shape that a hanging cable forms when it is supported at both ends is called a catenary and you can easily observe it if you hold the two ends of piece of thread and let it hang. It should look like a parabola, but in fact the shape it forms is not a parabola because $y = ax^2 + bx + c$ is not solution of the above differential equation.

To prove this, suppose $y = ax^2 + bx + c$ is a solution of the equation $\frac{y''}{\sqrt{1+(y')^2}} = k$.

Then $y' = 2ax + b$ and $y'' = 2a$ and $\frac{y''}{\sqrt{1+(y')^2}} = \frac{2a}{\sqrt{1+(2ax^2+b)^2}}$.

Therefore $\frac{y''}{\sqrt{1+(y')^2}} \neq k$ when $y = ax^2 + bx + c$.

> For simplicity, it is usual to write down derivatives using prime notation:
> $y^{(n)} = \frac{d^n y}{dx^n}$;
> $y^{(n-1)} = \frac{d^{n-1}y}{dx^{n-1}}$; ....; $y' = \frac{dy}{dx}$

In fact, catenaries and parabolas are different curves with different geometric properties and engineers need to know the difference between these shapes!

3.1 Classifications of differential equations and their solutions

The study of differential equations is a very important area of mathematics, but is also extremely vast. There are many different types of differential equations but no general method to solve all of them. For this reason mathematicians have a complex classification of differential equations that depends both on the form of the equation, and the methods known to solve it. Our study will focus on just a few simple cases: we will study three types of differential equation where the solution is of the form $y = f(x)$, where x is the independent variable and y is the dependent variable. The equations that relate a function of one variable with: its derivatives; an independent variable; and constants are called **Ordinary Differential Equations.**

> The study of differential equations is the area of mathematics that, more than any other, has been directly related to problems in mechanics, astronomy, and physics. Since the 17th century, when Newton, Leibniz, and the Bernoullis managed to solve some simple differential equations that arose from problems in geometry and mechanics, the creation of a huge 'bag of tricks' has not stopped. Although mathematicians have not been able to produce a general method to tackle all differential equations, some tricks actually work well and allow scientists to find answers to many of the problems they need to solve.

The main classification of ordinary differential equations (ODE) depends on the order of the derivatives that appear in the equation: the highest derivative order that appears in a differential equation is called **the order of the equation**. For example,

1 $\dfrac{dy}{dx} + y = 4$ is a first order differential equation

2 $\dfrac{d^2y}{dx^2} + 2\dfrac{dy}{dx} = y$ is a second order differential equation

3 $\dfrac{d^3y}{dx^2} + 5x\dfrac{dy}{dx} = e^x y$ is a third order differential equation

Equations (2) and (3) can be classified according to the type of coefficients: (2) has constant coefficients but in (3) the coefficients depend on x, so we say (3) is a differential equation with variable coefficients.

Example 1

Classify the following differential equations according to order and type of coefficients
a $y' - 4y = 5$ **b** $y''' + 5y'' - xy' = 4y + 3x$

a First order differential equation with constant coefficients.	*The coefficients are all constants. The highest order of the derivatives featured is 1.*
b Third order differential equation with variable coefficients.	*The coefficient of y' depends on x. The highest order of the derivatives featured is 3.*

Another important classification has to do with what is done with the function $y = y(x)$ and its derivatives: Ordinary Differential Equations can also be classified as linear or non-linear. For example, $xy' + \frac{2}{x}y = x^2$ and $\frac{2}{x}y' + 3xy = e^x$ are first order linear differential equations because they are of the form $a(x)\frac{dy}{dx} + b(x)\,y = c(x)$ where both the function ($y = y(x)$) and its derivative (y') are multiplied only by functions of x. Examples (2) and (3) above are second order linear equations.

A non-linear equation involves expressions such as y^2, $(y')^2$, $\sin(xy)$, e^y, ….

Here are a few examples of non-linear differential equations:

i $(y')^2 + 3y = 4$ is a first order non-linear differential equation;

ii $y' + 4x\sin(y) = 2$ is a first order non-linear differential equation;

iii $y'' + y' = e^{xy}$ is a second order non-linear differential equation.

What is a solution of a differential equation and how can we describe it?

Suppose a differential equation is of order n (this means the equation contains terms involving $y^{(n)}$, $y^{(n-1)}$, …, y'). A solution of this equation is a function $y = f(x)$ that can be differentiated at least n times in an interval I, and that when its derivatives ($y^{(n)}$, $y^{(n-1)}$, …, y') and the function itself (y) are substituted into the equation, both sides of the equation match for all values of x in the interval I. Sometimes this interval is the set of real numbers and we say that $y = f(x)$ *is a solution over the real numbers*, or *for all real numbers*.

Example 2

Verify that the function $f(x) = 2\sin(x) + 3\cos(x)$ is a solution of the linear differential equation $y'' + y = 0$ for all values of x.	
$f(x) = 2\sin(x) + 3\cos(x) \Rightarrow f'(x) = 2\cos(x) - 3\sin(x)$ $\Rightarrow f''(x) = -2\sin(x) - 3\cos(x)$	Use trigonometric differentiation rules twice
So $f''(x) + f(x) =$ $(-2\sin(x) - 3\cos(x)) + (2\sin(x) + 3\cos(x)) = 0$ for all values of x. We have therefore an identity and can conclude that $y = f(x)$ is indeed a solution of this ODE.	Substituting both y'' by $f''(x)$ and y by $f(x)$ into the expression $y'' + y$.

In the example above an explicit expression for $y = f(x)$ is given. However, often when solving differential equations it is not possible to obtain a solution in an explicit form, but only in implicit form.

Example 3

> Show that the equation $x^2 + y^2 = 1$ is an implicit solution of $\dfrac{dy}{dx} = \dfrac{xy}{x^2 - 1}$. State the values of x for which the function is defined.

$x^2 + y^2 = 1 \Rightarrow 2x + 2yy' = 0 \Rightarrow y' = -\dfrac{x}{y} \Rightarrow y' = -\dfrac{xy}{y^2}$	Differentiate implicitly and solve for y'
As $x^2 + y^2 = 1 \Rightarrow y^2 = 1 - x^2$,	
So $y' = -\dfrac{xy}{1-x^2}$ or $\dfrac{dy}{dx} = \dfrac{xy}{x^2-1}$.	Substituting $y^2 = 1 - x^2$ into the expression.
$\therefore x^2 + y^2 = 1$ is an implicit solution of $\dfrac{dy}{dx} = \dfrac{xy}{x^2-1}$.	This solution is defined for all values of $x \in \,]{-1}, 1[$.

Exercise 3A

1 For each of the following ordinary differential equations, state:
   a Whether they are linear or non-linear;
   b Their order;
   c Whether they have constant or variable coefficients.

Equation 1: $y^{(3)} = 6y$
Equation 2: $(y')^3 = 6y$
Equation 3: $y'' - e^x y' - y = 0$
Equation 4: $(y')^4 + 5y = 4$
Equation 5: $\sin(x)\, y' + 2x^2 y = \cos(x)$

Equation 6: $e^t \dfrac{dv}{dt} = kt^2$

Equation 7: $\dfrac{dv}{dt} = k\dfrac{t}{v}$

2 Verify whether or not each function $y = f(x)$ is a solution of the given differential equation.
   a $y' + y = 0$ and $y = e^{-x}$
   b $y'' + y = 0$ and $y = \sin(x)$
   c $y' + y = e^{-x}$ and $y = xe^{-x}$

3 Show that the following equations define implicit solutions of the given differential equations:

   a $e^{xy} + x + y = 0$ and $\dfrac{dy}{dx} = -\dfrac{1 + ye^{xy}}{1 + xe^{xy}}$

   b $x^2 + y^2 = r^2$ and $\dfrac{dy}{dx} = -\dfrac{x}{y}$

> The term *differential equation* was introduced by Leibniz in 1676. However, as you have probably noticed, *differential equations* are actually just equations which involve derivative terms, such as y' or y''. To this day, no serious attempt has yet been made to rename them *Derivative Equations*.
>
> Over the years, many famous mathematicians and scientists have dedicated themselves to the study of specific differential equations. Rather than change the general term for this type of equation, they simply gave their own name, or the name of the phenomenon they were studying, to these equations. For this reason, if you explore this topic you may come across terms like 'heat equation', 'Newton's laws', or 'Laplace's equations'.

Investigation: Do differential equations always have a solution? Can they have more than one solution?

1. Consider the differential equation $\frac{dy}{dx} - y = 0$. Show that $y = e^x$ and $y = 2e^x$ are solutions of this differential equation. Find a general expression for a family of functions that are solutions of this equation. Is there any function in this family whose graph contains the point $(0, 3)$?

2. Consider now the differential equation $\frac{dy}{dx} = y^2$. Try to find functions that are solutions of this differential equation. Do you know any function whose derivative squared is equal to the original function? If so, how many solutions can you find? Is there any solution whose graph contains the point $(1, 1)$?

3. Consider the differential equation $\left(\frac{dy}{dx}\right)^2 + 5 = 0$. Try to find functions that are solution of this differential equation. Do you know any function whose second derivative squared is equal to -5? Give reasons for your answer.

The previous investigation may have shown you that, in general, differential equations can have several solutions or even no solution. However, for practical applications, scientists who use differential equations to model a specific physical phenomenon need to consider additional conditions that allow them to find **the** solution to the problem they want to solve, rather than a whole family of solutions which would not be very helpful. These additional conditions are called **initial values** or **boundary conditions**. In general, as you learn how to solve specific types of differential equations, you first find general solutions (families of functions which satisfy the general type of differential equation you're looking at) and then you use the initial values or boundary conditions to determine the specific solution to your differential equation.

Example 4

a Show that $y = \ln(3x) + k$ is a general solution of the differential equation $y' = \dfrac{1}{x}$.

b Find the value of k and determine which function of the family $y = \ln(x) + k$ verifies the boundary condition $y(1) = 2$.

a $y' = \dfrac{3}{3x} \Rightarrow y' = \dfrac{1}{x}$	*Differentiate $y = \ln(3x) + k$*
b $2 = \ln(3) + k \Rightarrow k = 2 - \ln(3)$	*Substitute x by 1 and y by 2 to find the value of k*
$\therefore y = \ln(3x) + 2 - \ln(3) \Rightarrow y = \ln(x) + 2$ is the function of the family that satisfies the condition given.	

In this chapter we are going to study basic methods that will allow us to solve some special types of differential equations that are important due to their applications. These methods will also give a good indication of the skill and creativity employed by the pioneers of this area of mathematics as they tackled these problems.

3.2 Differential Equations with separated variables

A simple class of differential equations are called first order differential equations with separated variables. They are of the form $y' = f(x)$ and usually arise from problems about rates of change, where the rate of change y' only depends on the independent variable x. In most cases, these equations can be solved easily using the integration techniques studied in the core of Mathematics Higher Level, as shown in the examples below.

Example 5

Find the general solution of the differential equation $y' = x^2 + 1$. Hence find the solution that satisfies the initial condition $y(0) = 1$.

$y' = x^2 + 1 \Rightarrow y = \dfrac{x^3}{3} + x + c$	*Integrate both sides with respect to x.*
$y(0) = 1 \Rightarrow c = 1$	*Substitute x by 0 and y by 1.*
$\therefore y = \dfrac{x^3}{3} + x + 1$ satisfies the initial condition given.	*Substitute c by the value found.*

Example 6

Find the general solution of the differential equation $\frac{dx}{dt} = \cos(t)$. Graph a few solutions and comment on similarities between their graphs.

$\frac{dx}{dt} = -\sin(t) \Rightarrow x = \cos(t) + c$

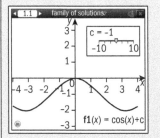
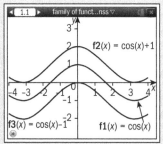

Each graph can be obtained from another one using a vertical translation.

By inspection find $x(t)$ such that $\frac{dx}{dt} = \cos(t)$.

Use a GDC to graph the solution for several values of c, or use sliders to visualize many graphs of this family of solutions.

Often we need to apply integration techniques like substitution or integration by parts to find a solution of a first order differential equation with separated variables.

Example 7

Find the particular solution of the differential equation that satisfies the initial condition given:

$\frac{dy}{dx} = \cos(x)\,e^{\sin(x)}$ and $y(0) = 0$.

Let $u = \sin x$

$\Rightarrow u' = \cos x \Rightarrow \frac{dy}{dx} = u'e^u$

$y = \int u'\,e^u\,dx = e^u + c$

$y = e^{\sin x} + c$

$y(0) = 0 \Rightarrow 0 = e^0 + c \Rightarrow c = -1$

$y = e^{\sin x} - 1$

Use integration by substitution

Substitute x by 0 and y by 0

Example 8

Find the general solution of the differential equation $\frac{dy}{dx} = x\sin(2x)$.

$y = \int x\sin(2x)\,dx = -\frac{x\cos(2x)}{2} + \frac{1}{2}\int \cos(2x)\,dx$

$\therefore y = -\frac{x\cos(2x)}{2} + \frac{1}{4}\sin(2x) + c$

Use integration by parts

Then write down the expression for the general solution

Modeling dynamic phenomena

Exercise 3B

1. Solve the following first order differential equations with separated variables. Check your answers using differentiation.

   a $\dfrac{dx}{dt} = t + t^2$ **b** $\dfrac{dy}{dx} = \dfrac{1}{x}$ **c** $\dfrac{dz}{dx} = x\cos(x)$ **d** $\dfrac{dw}{dx} = xe^{2x}$

2. Find the particular solution of the differential equation that satisfies the initial condition given:

   a $\dfrac{dy}{dx} = 3x + x^2$ and $y(1) = 4$ **b** $\dfrac{dy}{dx} = \dfrac{-x}{\sqrt{4-x^2}}$ and $y(0) = 1$

3. A function f is a solution of the differential equation given by $\dfrac{dy}{dx} = \dfrac{1}{x+2} - \dfrac{1}{2}\sin(x)$ for $x \geq -1$. The graph of f passes through the point $(0, 2)$. Find an expression for $f(x)$.

4. Show that $y = \sin(kx) - kx\cos(kx)$ (k is a constant) is a solution of the first order differential equation $\dfrac{dy}{dx} = k^2 x \sin(kx)$

5. Find the solution of the differential equation $\dfrac{dy}{dx} = e^{-2x} - \dfrac{1}{x-1}$, $x < 1$ that satisfies the initial condition $y(0) = 1$

6. A particle is projected along a straight line path. After t seconds, its velocity v metres per second is given by $v = \dfrac{1}{2+t^2}$. Find the distance travelled by the particle in the first t seconds.

3.3 Separable variables, differential equations and graphs of their solutions

In this section we will explore a variety of problems that arise when studying rates of change. Many of these problems can be modelled by first order differential equations with separable variables. These equations can be written in the form

$$\dfrac{dy}{dx} = \dfrac{f(x)}{g(y)}$$

> When $g(y) = 1$ we have the particular case studied in **3.2**

To solve these equations we just need to re-arrange them as $g(y)\dfrac{dy}{dx} = f(x)$ and then recall the chain rule and the fact that $y = y(x)$. In order to integrate both sides of the equation with respect to x, we can rearrange the equation as:

$$g(y(x))\dfrac{dy}{dx}dx = f(x)dx$$

Integrating both sides of the equation with respect to x, you obtain an implicit general solution of the equation:

$$\int g(y)\, dy = \int f(x)\, dx + c.$$

Sometimes it is possible to solve this equation for y and, provided a boundary condition is given, determine the particular solution that satisfies it.

Let's look at a few examples to clarify the process:

Example 9

Solve the following separable differential equation $\dfrac{dy}{dx} = -\dfrac{x}{y}$	
$\dfrac{dy}{dx} = -\dfrac{x}{y} \Rightarrow y\, dy = -x\, dx$	Separate the variables.
$\int y\, dy = -\int x\, dx \Rightarrow \dfrac{y^2}{2} + \dfrac{x^2}{2} = c$ Re-writing this, $x^2 + y^2 = r^2$ So the solutions of this equation are circles centred at origin and radii r.	Integrate both sides, and re-arrange the equation. The solution can be written as $x^2 + y^2 = r^2$ with $r \geq 0$. The radius is determined when boundary conditions are given. For example, if we required $y(0) = 1$, we would obtain $r = 1$.

It is also possible to first analyse the differential equations graphically, and this way predict the graph of the solutions. These graph displays of differential equations are called **slope fields** and, for each point on the grid, they show the gradient to the curve described by the solution $y = y(x)$. For example, the slope field for

$\dfrac{dy}{dx} = -\dfrac{x}{y}$ is shown on the right:

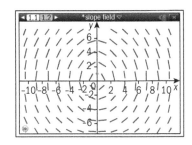

If we look at the solution found in the previous example we can recognise its circular graph being obtained from the slope field we showed above. Remember that each little line in the slope field represents the tangent to the curves of $x^2 + y^2 = r^2$ at that point. Can you see the graph of the solution that contains the point $(0, 1)$?

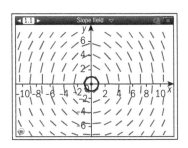

For these simple cases they can in fact be plotted manually. For example, if we consider now the differential equation $\frac{dy}{dx} = \frac{x}{y}$ we can easily plot the slope field by hand. Start with a two-entry table where you record the value of the derivative at (x, y):

> Slope fields are also called phase graphs or phase portraits.

	0	1	2	3	4	5
0	nd	nd	nd	nd	nd	nd
1	0	1	2	3	4	5
2	0	0.5	1	1.5	2	2.5
3	0	0.333…	0.666…	1	1.33…	1.666
4	0	0.25	0.5	0.75	1	1.25
5	0	0.2	0.4	0.6	0.8	1

For each point (x, y) draw a short line segment with the slope shown in the table:

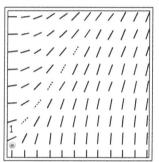

> 'nd' means 'no derivative at this point'

We can confirm our results and extend the slope field for other values of (x, y):

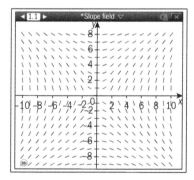

Isoclines and graphical approximations of solutions

The term isocline derives from the Greek for "same slope."
For a first-order differential equation $\frac{dy}{dx} = f(x, y)$, the curve with equation $f(x, y) = m$ for some constant m is known as an isocline. This means that all the solutions ($y(x)$) of the ordinary differential equation $\frac{dy}{dx} = f(x, y)$ that intersect the curve $f(x, y) = m$ have the same slope, m.

Isoclines can be used as a graphical method of solving differential equations.

Let's solve the differential equation $\frac{dy}{dx} = \frac{x}{y}$, and then draw in the isoclines to illustrate how we can use isoclines to graphically solve differential equations.

$$\frac{dy}{dx} = \frac{x}{y} \Rightarrow y\,dy = x\,dx \Rightarrow \int y\,dy = \int x\,dx \Rightarrow \frac{y^2}{2} - \frac{x^2}{2} = c \text{ or } y^2 - x^2 = k.$$

The solutions are curves called equilateral hyperbolas with centre at the origin. The isoclines are the lines $y = \frac{1}{m}x$. Note that for each value of m the isocline intersects the curves (graphs of the solutions) at points where the tangents to their graphs are parallel.

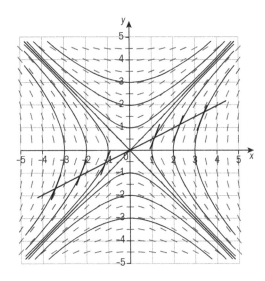

> Software like Geogebra allows you to plot slope field, graph your guess at the solution and show isoclines.

> Isoclines provide a quick method to sketch slope fields: First draw the isoclines, and then add line-segments which cross the isocline and are all of equal gradient. These line-segments represent the tangents to the solutions of the differential equation at that point. Note that for each isocline $f(x, y) = m$, the slope of the isocline corresponds to the value of m.

Example 10

Solve the following separable differential equation $\frac{dy}{dx} = y + 1$	
$\frac{dy}{dx} = y + 1$	Separate the variables.
$\int \frac{1}{y+1} dy = \int dx$	Integrate both sides.
So a general solution of the equation is $\ln\|y+1\| = x + c \Rightarrow y + 1 = \pm e^{x+c} \Rightarrow y = Ae^{x-1}$.	Solve for y if you want to obtain the general solution explicitly, i.e. in the form $y = f(x)$

Modeling dynamic phenomena

We can visualize the solutions of the differential equation above by plotting the isoclines $y + 1 = m$ for different values of m and sketch the corresponding slope field. For example,

$m = 0 \Rightarrow y = -1$, $m = 1 \Rightarrow y = 0$, $m = -1 \Rightarrow y = -2$, and $m = 2 \Rightarrow y = 1$.

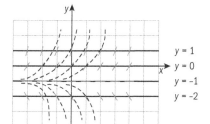

Example 11

Solve the following separable differential equation $\dfrac{dy}{dx} = \dfrac{x}{\sin(y)e^x}$	
$\dfrac{dy}{dx} = \dfrac{x}{\sin(y)e^x}$	Separate the variables.
$\displaystyle\int \sin(y)\,dy = \int xe^{-x}\,dx$	Integrate both sides. Using integration by parts, $\displaystyle\int xe^{-x}\,dx = -xe^{-x} + \int e^{-x}\,dx = -(x+1)e^{-x} + k$
So a general implicit solution of the equation is $\cos(y) = (x + 1)e^{-x} + c$.	

In the example above, explicit solutions can be obtained if initial conditions are set. For example, if $y(0) = 2\pi \Rightarrow c = 0$ and a particular solution is $\cos(y) = (x + 1)e^{-x} \Rightarrow y = \arccos((x + 1)e^{-x}) + 2\pi$. But if $y(0) = 4\pi \Rightarrow c = 0$ and a particular solution is $\cos(y) = (x + 1)e^{-x} \Rightarrow y = \arccos((x + 1)e^{-x}) + 4\pi$.

This example shows that we may have the same implicit solution that leads to distinct explicit solutions, depending on the initial solutions set. To understand the situation better it is useful to graph the implicit equation: $\cos(y) = (x + 1)e^{-x}$

> For $c = 0$, the implicit equation defines a family of curves with equation
> $y = \arccos((x+1)e^{-x}) + 2k\pi$, $k \in \mathbb{Z}$

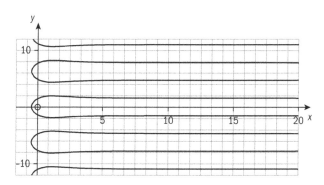

We can see that there are many explicit solutions of the same shape, but they sit in different positions relative to the y-axis. They all stem from the same implicit solution.

Exercise 3C

1 Consider the variable separable equations of the form $\frac{dy}{dx} = f(y)$.

Re-write each equation in the form $\frac{dy}{f(y)} = dx$ and solve it.

You may leave your answers in an implicit form for y.

> You may need to use integration by substitution to find the solution to some of these differential equations.

a $\frac{dy}{dx} = y - 3$ **b** $\frac{dy}{dx} = \frac{1}{3y^2 + y - 1}$ **c** $\frac{dy}{dx} = y^2 + 4$

d $\frac{dy}{dx} = 2y + 1$ **e** $\frac{dy}{dx} = \cos^2(3y - 1)$ **f** $\frac{dy}{dx} = \frac{1 + \sin^2 y}{\cos y}$

2 Separate the variables and then solve the following differential equations. Check your answers by using differentiation. Give your answers as expressions for y in terms of x.

a $e^{-x} \frac{dy}{dx} = \frac{3}{y}$ **b** $\frac{5}{y-1} \frac{dy}{dx} = \frac{2}{x}$ **c** $\frac{dy}{dx} = \frac{y}{x}$

d $\frac{dy}{dx} = -\frac{y^2 + 1}{x^2 + 1}$ **e** $e^{-x} \frac{dy}{dx} = \frac{2x}{y}$ **f** $\frac{x}{y^2 + 1} \frac{dy}{dx} = \frac{2}{x}$

3 Solve the following initial value problems:

a $(1 + \cos(2x)) \frac{dy}{dx} = 2\sin(2x) y$, and $y(0) = 1$

b $(1 + x^2) \frac{dy}{dx} = 1 + y^2$, and $y(0) = 2$

c $e^x \frac{dy}{dx} + 1 = y$, and $y(0) = 2$

4 Air is pumped into a spherical ball, with radius $r = r(t)$, which expands at a rate of $8\,\text{cm}^3$ per second.
  a Write down a differential equation of the form $\frac{dr}{dt} = f(t)$ that models this situation.
  b Solve the equation and determine an expression for the radius of the ball in terms of t, given that $r(1) = 5$

5 Consider the differential equation $\frac{dy}{dx} = e^{-xy}$.
  a State whether or not this differential equation is a variable separable equation.
  b Use a GDC or computer software to sketch the slope field for this differential equation.
  c Estimate graphically the curve solution to the differential equation through the origin.

Modeling dynamic phenomena

6 Consider the first order differential equations $\frac{dy}{dx} = \frac{2y}{x}$ and $\frac{dy}{dx} = -\frac{x}{2y}$.

   a For each equation draw the isoclines $\frac{2y}{x} = m$ and $-\frac{x}{2y} = m$, for $m = 0, \pm 1, \pm 2, \pm 4$.

   b Hence draw the slope fields for each equation.

   c Sketch the graphs of possible solutions for each equation and describe them geometrically.

   d Use the expressions of the differential equations given to show that the graphs of the solutions to each equation are orthogonal at the point where they intersect.

7 Use isoclines to obtain the slope fields for each of the following equations. Then use technology to confirm your answers.

   a $\frac{dy}{dx} = xy$

   b $\frac{dy}{dx} = \frac{y}{x^2}$

3.4 Modeling of growth and decay phenomena

An important class of separable differential equation has the form $\frac{dy}{dt} = \pm ky$ where $k > 0$ and t is the independent variable *time*. These equations can be solved using separation of variables:

$$\frac{dy}{dt} = \pm ky \Rightarrow \frac{dy}{y} = \pm k\,dt \Rightarrow \ln|y| = \pm kt + c \Rightarrow |y| = e^{\pm kt + c} \Rightarrow y = Ae^{\pm kt}$$

where A is a positive constant to be determined using the initial or boundary conditions.

The general solution $y = Ae^{kt}$ is called the **exponential growth curve** which provides a model for many real life growth and decay problems.

For example, we can use this model to represent how to calculate continuous compound interest:

If a bank that pays the annual interest of r, compounded continuously, then the future value $A(t)$ of the account satisfies the conditions:

$\frac{dA}{dt} = rA$ and $A(0) = A_0$ where A_0 is the initial amount deposited in the bank.

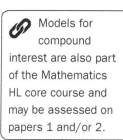
Models for compound interest are also part of the Mathematics HL core course and may be assessed on papers 1 and/or 2.

The solution of this equation is an exponential function: $A(t) = A_0 e^{rt}$.

In the investigation below we will study the relation between the continuous model and the discrete models of compound interest.

Short Investigation

Suppose that you decide to invest 1000 euros (or any other currency you prefer) for a period of 40 years. The bank offers you an interest rate of 3.6% p.a. but you are not told how this is going to be calculated. Create a table to analyse the future value of your money if the interest is calculated annually, quarterly, monthly, daily, hourly, or continuously. Use line graphs to show your results. Plot each line graph on the same grid; considering $t \in [0, 40]$ (t in years). Compare the graphs and comment on the differences you observe.

> Before the 1940s, scientists had no accurate way of determining the age of fossils or other ancient objects. They had to rely on relative dating techniques, which typically held great potential for error. In 1948 the scientist Willard Frank Libby came up with the method known as radiocarbon dating: Plants and animals which utilise carbon in biological foodchains take up C^{14} during their lifetimes. They exist in equilibrium with the C^{14} concentration of the atmosphere. This means that the proportion of C^{14} atoms to non-radioactive carbon atoms within the organism is approximately the same as that in the atmosphere around, during the time that the organism is alive. As soon as a plant or animal dies, they cease the metabolic function of carbon uptake; there is no replenishment of radioactive carbon, but only decay. He found that after 5568 years, half the C^{14} in the original sample will have decayed and after another 5568 years, half of that remaining material will have decayed, and so on. This contribution to development of science made him the winner of the Nobel Prize for Chemistry in 1960.

The general solution to $\dfrac{dy}{dt} = -ky$ is $y = Ae^{-kt}$. Its graph is called the **exponential decay curve** which is a useful model to represent phenomena like radiocarbon dating.

> Some GDCs allow us to vizualize the slope fields of exponential growth and decay easily. Adding a slider to vary the value of the constant, k, allows us to observe the change in shape of the slope fields as we move from negative to positive values.

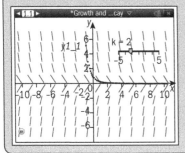
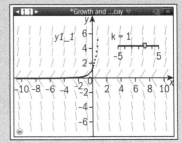

Modeling dynamic phenomena

Example 12

The **Lascaux Caves** in southwest France contain some of the oldest and finest prehistoric art in the world. By means of complex chemical analysis on charcoal taken from the caves, scientists were able to determine that the charcoal contained 15% of the original amount of C^{14} that it would have contained when the tree it was made from was cut down. Using Libby's half-life model for C^{14} described above,

a Find the value of the decay constant k for the model of that gives the quantity Q of C^{14} present in the sample of charcoal.

b Hence find an approximate value for the age of Lascaux paintings.

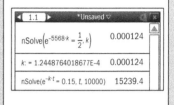

You may use your calculator solver to find the answers but you may need to enter a wise guess to obtain the solution

a The exponential decay curve is given by $Q(t) = Q_0 e^{-kt}$, where Q_0 is the initial quantity of C^{14}.

$$Q(5568) = Q_0 e^{-5568k} = \frac{1}{2} Q_0$$

$$\Rightarrow k = \frac{\ln 2}{5568} \Rightarrow k = 0.000124 \text{ (3 s.f.)}.$$

b $Q_0 e^{-kt} = 0.15 Q_0 \Rightarrow t = -\frac{\ln(0.15)}{k}$

$\Rightarrow t = 15200$ years (3 s.f.)

Assuming that the model is described by

$$\frac{dQ}{dt} = -kQ \text{ and } Q(0) = Q_0$$

Substitute half-life value into decay curve model expression and solve for k.

Solve the equation $Q(t) = 0.15 Q(0)$

Exercise 3D

1 By chemically analysing the remains of a cypress beam found in an archaeological site in Egypt (the Tomb of Sneferu) scientists estimated that the amount of C^{14} present in the sample was approximately 55% of the amount that is found in a living cypress tree. Use this information and Libby's half-life model to estimate the age of the tomb.

2 Huize read that some scientists disagree with Libby's estimation of the value of the half-life value for C^{14}, and claim that a more accurate estimation is 5530 years. She decides to use this value to obtain another estimation of the age of the tomb described in question 1. Find the new estimation value.

3 Assuming that the rate at which alcohol is eliminated from the blood circulation is proportional to the concentration of alcohol in the blood of an individual,
   a Calculate an equation for the content ($V(t)$) of alcohol in a person's blood after t hours of consuming the alcohol, assuming that the rate at which the alcohol is eliminated is 0.015% per hour, and the initial blood content of alcohol is V_0.
   b Yong read a newspaper article which claimed that "… two hours after an accident, test results indicated that a driver had a blood alcohol content of 0.08%". Yong is wondering if the driver's blood alcohol rate was within the permissible limits at the time of the accident. If the permissible level for drivers is 0.10%, use the model to decide whether or not the driver's blood alcohol rate was within the permissible limits at the time of the accident.

4 The rate of growth of a population, P, of bears in a national park can be modelled by the differential equation

$$\frac{dP}{dt} = 0.0004P(250-P),$$ where t is time in years.

$$\frac{1}{P(250-P)} = \frac{1}{250}\left(\frac{1}{P} + \frac{1}{250-P}\right)$$

Find an expression for the population P of bears in terms of t, given that $P(0) = 35$. Hence determine the number of years it will take until the population reaches 240 bears.

5 Artjom has been exploring economic models for inflation. He decides to consider a model where the inflation decreases over time according to the model $\frac{dP}{dt} = \frac{I_0}{1+t}P$ where t is in years, $P(t)$ is the price of an item at a time t, and I_0 is the initial inflation rate.
   a Assuming that the inflation rate starts at 3%, calculate the price of an item in 10 years if $P(0) = 100$ euros.
   b Find an expression for $P(10)$ in terms of I_0 given that $P(0) = 100$ euros.

   John thinks that actually the inflation will increase in the near future and suggests a different model: $\frac{dP}{dt} = (0.01 + 0.001t)P$.
   c Solve the differential equation and find the value of $P(10)$ given that $P(0) = 100$ euros.

EXTENSION QUESTION

6 Emil and Kevin decided to challenge themselves and tackle a problem about terminal velocity of a falling object. They know that the velocity of an object of mass m falling from rest under the action of gravity is modelled by the differential equation $m\frac{dv}{dt} = mg - kv^2$ where g is the acceleration due to gravity and k is a constant that depends on the both the aerodynamic properties of the object, and density of the air.
   a Let $\alpha = \sqrt{\frac{gk}{m}}$. Show that the differential equation given can be written as $\left(\frac{1}{g-\alpha v} + \frac{1}{g+\alpha v}\right)dv = 2dt$.
   b Hence solve the differential equation and find $\lim_{t\to\infty} v(t)$.

3.5 First order exact equations and integrating factors

Many of the advances in mathematics have a touch of human ingenuity! This was certainly the case for mathematicians *Johann Bernoulli* and *Leonhard Euler*, both foundational figures in the development of the calculus and of the theory of differential equations.

> 🔍 The Swiss mathematician Johann Bernoulli (1667–1748) was one of the pioneers in the field of calculus and helped apply this new tool to real-world problems. He was a member of one of the world's most successful mathematical families, the Bernoullis. Johann himself did important work on the study of the equation $y = x^x$, the study of the Bernoulli series, and made advances in theory of navigation and ship sailing. In addition, he is famous for his tremendous letter-writing (he wrote over 2500 letters), and his tutoring of another great mathematician, Leonhard Euler. As for the rest of the Bernoulli family, there were eight good mathematicians over three generations.

In this section we are going to study a clever method that will allow us to solve equations that can be written in the form $\frac{dy}{dx} + p(x)y = q(x)$. First, however, let's look at a special case of first order equations called **exact equations.** The equation $x^2 \frac{dy}{dx} + 2xy = 1$ is an example of an exact equation. This equation is clearly a linear differential equation as it can be written as $\frac{dy}{dx} + \frac{2}{x}y = \frac{1}{x^2}$, for $x \neq 0$. However if you observe it in its original form you may notice that $x^2 \frac{dy}{dx} + 2xy = \frac{d}{dx}(x^2 y)$ which means that the equation can be rewritten as $\frac{d}{dx}(x^2 y) = 1$ and solved by integrating each side with respect to x:

$$x^2 y = x + c \Rightarrow y = \frac{c}{x^2} + \frac{1}{x}, x \neq 0.$$

Equations like this one, that can be written in the form
$\frac{d}{dx}(u(x) \cdot y) = v(x)$ for some function $u(x)$, are called **first order exact equations**. Most first order equations are not exact, but can be transformed into exact equations by multiplying both sides of the equation by an appropriate expression called an **integrating factor**.

For example, $xy\frac{dy}{dx} + y^2 = 3x$ is not exact, but if you multiply both sides by $2x$ you obtain $2x^2 y \frac{dy}{dx} + 2xy^2 = 6x^2$, or $\frac{d}{dx}(x^2 y^2) = 6x^2$, which is exact. If we can find an integrating factor that makes a first order equation exact then the problem of solving it becomes simply an integration problem.

Fortunately, for first order linear equations of the form $\frac{dy}{dx} + p(x)y = q(x)$, we can find an integrating factor $I(x)$ in a systematic way as stated in the theorem below:

Theorem 1: Given the differential equation $\frac{dy}{dx} + p(x)y = q(x)$, the function $I(x) = e^{\int p(x)dx}$ is an integrating factor that transforms this differential equation into an exact differential equation of the form

$$\frac{d}{dx}\underbrace{\left(e^{\int p(x)dx} y\right)}_{u(x)} = \underbrace{e^{\int p(x)dx} q(x)}_{v(x)}.$$

> When calculating the integrating factor, we usually take the integration constant equal to zero. You may want to investigate why the value of this constant is not important.

Proof: Multiply both sides of the equation by $I(x)$:

$$e^{\int p(x)dx} \frac{dy}{dx} + p(x) \cdot e^{\int p(x)dx} y = e^{\int p(x)dx} q(x)$$

As

$$\frac{d}{dx}\left(e^{\int p(x)dx}\right) = \left(\frac{d}{dx}\left(\int p(x)dx\right)\right) \cdot e^{\int p(x)dx} = p(x) \cdot e^{\int p(x)dx}$$

and

$$e^{\int p(x)dx} \frac{dy}{dx} + p(x) \cdot e^{\int p(x)dx} y = \frac{d}{dx}\left(e^{\int p(x)dx} y\right)$$

then the equation can be written as

$$\frac{d}{dx}\left(e^{\int p(x)dx} y\right) = e^{\int p(x)dx} q(x).$$

Modeling dynamic phenomena

Example 13

Consider the first order linear equation $x\dfrac{dy}{dx}+3y=e^{x^3}$.

a Find an integrating factor for this differential equation.
b Hence solve the differential equation.

---

a $x\dfrac{dy}{dx}+3y=e^{x^3} \Rightarrow \dfrac{dy}{dx}+\dfrac{3}{x}y=\dfrac{e^{x^3}}{x}, \; x\neq 0$ | Write it in the form $\dfrac{dy}{dx}+p(x)y=q(x)$

$I(x)=e^{\int \frac{3}{x}dx}=e^{3\ln x}=x^3$ (for simplicity $C=0$) | $I(x)=e^{\int p(x)dx}$ where $p(x)=\dfrac{3}{x}$

b $x^3\cdot\left(\dfrac{dy}{dx}+\dfrac{3}{x}y\right)=x^3\cdot\left(\dfrac{e^{x^3}}{x}\right) \Rightarrow x^3\dfrac{dy}{dx}+3x^2y=x^2e^{x^3}$ | Multiply both sides by $I(x)$

$\dfrac{d}{dx}(x^3y)=x^2e^{x^3} \Rightarrow x^3y=\dfrac{1}{3}e^{x^3}+c$ | Integrate both sides

$y=\dfrac{1}{3x^3}e^{x^3}+\dfrac{c}{x^3}, \; x\neq 0$ | Solve for y

Example 14

Find the particular solution of the following first order linear differential equation that satisfies the initial condition, given $\dfrac{dy}{dx}+2y=e^x$ and $y(0)=1$

$\dfrac{dy}{dx}+2y=e^x \Rightarrow I(x)=e^{\int 2dx}=e^{2x}$ | $I(x)=e^{\int p(x)dx}$ where $p(x)=2$

$e^{2x}\cdot\left(\dfrac{dy}{dx}+2y\right)=e^{2x}\cdot(e^x) \Rightarrow e^{2x}\dfrac{dy}{dx}+2e^{2x}y=e^{3x}$ | Multiply both sides by $I(x)$

$\dfrac{d}{dx}(e^{2x}y)=e^{3x} \Rightarrow e^{2x}y=\dfrac{1}{3}e^{3x}+c$ | Integrate both sides

$y=\dfrac{1}{3}e^x+\dfrac{c}{e^{2x}}$ or $y=\dfrac{1}{3}e^x+ce^{-2x}$ | Solve for y

Exercise 3E

1 Solve the following first order linear equations using an appropriate integrating factor:

a $\dfrac{dy}{dx}+3y=e^{2x}$

b $\dfrac{dy}{dx}+(2x+1)y=e^{-x^2}$

c $\dfrac{dy}{dx}+y=e^{-2x}\cos(x)$

d $\dfrac{dy}{dx}+y\tan(x)=\cos(x)$

e $\dfrac{dy}{dx}+y\cot(x)=\cos(x)$

2 Find particular solutions to the following first order linear equations that satisfy the initial condition given.

a $\dfrac{dy}{dx} = x - y$ and $y(0) = -2$ **b** $\dfrac{dy}{dx} - 2y = \sin x$ and $y(0) = 0$

c $\dfrac{dy}{dx} + (x+1)y = e^{-\frac{x^2}{2}}$ and $y(0) = 2$

3 Show that the following equations are exact, and hence solve them. In each case, say whether or not the equation is a first order linear equation.

a $x\dfrac{dy}{dx} + y = x^2 + x$ **b** $(x+1)\dfrac{dy}{dx} + y = \sin x \, e^x$

c $(\cos x)\dfrac{dy}{dx} - (\sin x)y = \cos^4 x$ **d** $(\sin x)\dfrac{dy}{dx} + (\cos x)y = \tan x$

4 Show that the following equations are both first order linear equations, and also separable variable equations. Hence solve each of them using two different methods.

a $x^2\dfrac{dy}{dx} - x^2 y = y$ **b** $(x^2+1)\dfrac{dy}{dx} - xy = 0$ **c** $\dfrac{1}{x}\dfrac{dy}{dx} + 2y = 3$

EXTENSION QUESTION

5 Show that the substitution $u = \dfrac{1}{y}$ transforms the first order non-linear equation $\dfrac{dy}{dx} + xy(1-y) = 0$ into a linear equation of the form $\dfrac{du}{dx} + p(x)u = q(u)$. Hence solve the equation and find the particular solution that satisfies the initial condition $y(1) = 2$

Real life applications of first order linear differential equations

Newton's cooling law

The mathematical formulation of Newton's empirical law of cooling of an object is given by the first-order linear differential equation $\dfrac{dT}{dt} = \alpha(T - T_e)$ where T represents the temperature of the object at a time t, and T_e represents the temperature of the environment surrounding the object (e.g. the temperature of the room where the object is. We assume the temperature of the environment remains constant). This equation is both a first order linear equation (since $\dfrac{dT}{dt} = \alpha(T - T_e) \Rightarrow \dfrac{dT}{dt} - \alpha T = -\alpha T_e$), and also a separable variables equation. We can therefore solve it using two different methods and obtain the cooling model equation $T(t) = ke^{\alpha t} + T_e$.

> These applications illustrate the type of contexts that may appear in exam questions. You are not required to know any of these formulas but you may be guided to deduce and apply some of them.

Modeling dynamic phenomena

Example 15

The temperature of a kitchen is 25°C. When a chicken is removed from an oven, its temperature is 175°C. Three minutes later, the temperature of the chicken has decreased to 155°C. Determine how many minutes it will it take for the chicken to cool off to a temperature of 50°C.

$T(t) = ke^{at} + T_e$ where $T(0) = 175$, $T(3) = 155$ and $T_e = 25$	Using Newton's cooling law
$T(0) = 175 \Rightarrow ke^0 + 25 = 175 \Rightarrow k = 150$	Solve $T(0) = 175$ and $T(3) = 155$ simultaneously
$T(3) = 155 \Rightarrow 150e^{3\alpha} + 25 = 155 \Rightarrow \alpha = \dfrac{1}{3}\ln\left(\dfrac{13}{15}\right)$	
Then $150e^{\frac{1}{3}\ln\left(\frac{13}{15}\right)t} + 25 = 50 \Rightarrow t = 38$ (to the nearest minute)	Solve $T(t) = 50$ for t in minutes. 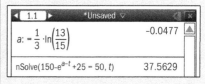

Falling body and Newton's law

Suppose that a body with mass m is dropped from rest from a great height in the Earth's atmosphere. To simplify the model, let's assume that: the body falls in a straight line; that the acceleration g due to gravity is constant; and that the air resistance force is proportional to its velocity, v.

Under these conditions, Newton's second law gives you the following equation

$$m\frac{dv}{dt} = mg - kv$$

with initial conditions $v(0) = 0$.

Example 16

Show that a general solution of $m\dfrac{dv}{dt} = mg - kv$ when $v(0) = 0$ is given by $v(t) = \dfrac{mg}{k}\left(1 - e^{-\frac{k}{m}t}\right)$.

$m\dfrac{dv}{dt} = mg - kv \Rightarrow \dfrac{dv}{dt} + \dfrac{k}{m}v = g \quad I(t) = e^{\frac{k}{m}t}$	Write it in the form $\dfrac{dy}{dx} + p(x)y = q(x)$. Calculate the integrating factor.
$e^{\frac{k}{m}t}\dfrac{dv}{dt} + \dfrac{k}{m}e^{\frac{k}{m}t}v = g\,e^{\frac{k}{m}t}$	Multiply both sides of the equation by $I(t)$ to make it exact.
$e^{\frac{k}{m}t}v = \displaystyle\int g\,e^{\frac{k}{m}t}\,dt$	
$e^{\frac{k}{m}t}v = \dfrac{gm}{k}e^{\frac{k}{m}t} + c \Rightarrow v = \dfrac{gm}{k} + c\,e^{-\frac{k}{m}t}$	Integrate both sides with respect to t and re-arrange for v.
$v(0) = 0 \Rightarrow 0 = \dfrac{gm}{k} + c.e^0 \Rightarrow c = -\dfrac{gm}{k}$	Use initial condition to find the value of c.
$\therefore v = \dfrac{gm}{k} - \dfrac{gm}{k}e^{-\frac{k}{m}t} \Rightarrow v = \dfrac{mg}{k}\left(1 - e^{-\frac{k}{m}t}\right)$	Substitute the value of c into the equation and re-arrange it to required form.

Mixing solutions problems

In chemistry and biology, scientists often need to predict the amount of certain substances (e.g. drugs, salt, or hormone) in a container in which a liquid mixture of the substance is continuously being added or extracted.

This type of problem is generally represented by a first order linear differential equation of the form $\frac{dQ}{dt} + p(t)Q = q(t)$ where $p = p(t)$ represents the rate at which the liquid is exiting the container, and $q = q(t)$ represents the rate at which liquid is entering the container.

Example 17

Fahmi is doing a chemical experiment. He fills a tank with 100 litres of pure water. A salt solution containing 20g of salt per litre of solution is poured into the container at a rate of 10 litres per minute. The mixture is continuously stirred and drained out of the tank at the same rate. Estimate the quantity Q of salt in the tank 20 minutes after the start of the experiment.

$\frac{dQ}{dt} = 20 \times 10 - \frac{Q}{100} \times 10 \Rightarrow \frac{dQ}{dt} + \frac{Q}{10} = 200$	Rate in = (concentration in) (flow rate in) Rate out = (concentration out) (flow rate out) where concentration is expressed in grams per litre
We have 100 because the flow rate in is equal to the flow rate out and so there will always be 100 litres of solution in the tank.	
$I(t) = e^{\frac{1}{10}t}$	
$e^{\frac{1}{10}t}\frac{dQ}{dt} + \frac{Q}{10}e^{\frac{1}{10}t} = 200e^{\frac{1}{10}t}$	Calculate the integrating factor and multiply both sides of the equation by I to make the equation exact.
$\Rightarrow e^{\frac{1}{10}t}Q = 2000e^{\frac{1}{10}t} + c$	Integrate both sides with respect to t and re-arrange for Q.
$\Rightarrow Q = 2000 + c\, e^{\frac{-1}{10}t}$	Use initial condition to find the value of c.
$Q(0) = 0 \Rightarrow c = -2000$	Substitute the value of c to obtain the solution of the differential equation that satisfies the initial conditions.
$\therefore Q = 2000 - 2000\, e^{-\frac{1}{10}t}$	
$Q(20) = 2000\left(1 - e^{-\frac{1}{10}\times 20}\right) = 1730$ grams (3 s.f.)	Find Q, in grams, when $t = 20$ minutes.

Continuous compounding interest revisited

When an initial amount A_0 is deposited in a bank that pays a total annual interest of r (represented in decimal form) and if deposits amounting a total of D (in the same currency) are made throughout the year (for simplicity, assume that they are made continuously), then the value $A = A(t)$ of the account satisfies the linear differential equation $\frac{dA}{dt} = rA + D$ with initial condition $A(0) = A_0$.

Example 18

Solve the initial value problem $\frac{dA}{dt} = rA + D$, with initial condition $A(0) = A_0$.	
$\frac{dA}{dt} = rA + D \Rightarrow \frac{dA}{dt} - rA = D$	Re-arrange into the form $\frac{dy}{dx} + p(x)y = q(x)$ and calculate the integrating factor.
$I(t) = e^{\int -r dt} = e^{-rt}$	
$e^{-rt}\frac{dA}{dt} - r e^{-rt} A = D e^{-rt} \Rightarrow \frac{d}{dt}(e^{-rt} A) = D e^{-rt}$	Multiply both sides of the equation by the integrating factor to make the equation exact.
$e^{-rt} A = D \int e^{-rt} dt \Rightarrow e^{-rt} A = -\frac{D}{r}e^{-rt} + c$	Integrate both sides with respect to t.
$A(0) = A_0 \Rightarrow A_0 = -\frac{D}{r} + c \Rightarrow c = A_0 + \frac{D}{r}$	Use initial condition to find the value of c.
$\therefore A = -\frac{D}{r} + \left(A_0 + \frac{D}{r}\right)e^{rt}$ or $A = A_0 e^{rt} + \frac{D}{r}(e^{rt} - 1)$	Substitute the value of c to obtain the solution of the differential equation that satisfies the initial conditions. Re-arrange for A to obtain an explicit formula for A in terms of t.

Exercise 3F

1 A thermometer that has been in a room for a long time shows the temperature of the room is 24°C. Te-Chen takes it outside and, five minutes later, the thermometer reads 22°C. After another 5 minutes, it reads 20.4°C. Estimate the outside temperature, assuming that it remains constant for the next few hours.

> Use Newton's Law of Cooling

2 Upon his graduation, Hamdi decides to start a savings plan: he starts with $1000 and gets an amazing deal that will pay him a fixed interest rate of 4% for the next 50 years. Every year Hamdi is planning to add an extra $5000 to his savings fund.
   a Assuming that the interest is calculated continuously, write down the values of the constants a and b that make the following differential equation a model of the rate of change of the amount x available in Hamdi's fund after t years: $\frac{dx}{dt} = a + bx$
   b State the initial condition and solve the equation.
   c How many years it will take for Hamdi's fund to reach the value of $100 000? Will Hamdi's fund reach $1 000 000 in the next 50 years?

3. A skydiver, fully equipped, has a mass of 90 kg. He jumps out of an airplane at a height of 750 metres. The air resistance, k, is assumed to be $k = 0.5$ before the parachute opens, and $k = 10$ when the parachute opens. The parachute opens after 10 seconds. Find:
   a the velocity of the skydiver when the parachute opens;
   b the distance the skydiver has already travelled at the instant the parachute opens;
   c the limiting velocity the skydiver can reach after the parachute is open.

4. Huize is doing a chemical experiment. She fills a tank with 100 litres of pure water. A salt solution containing 10 g of salt per litre of solution is poured into the container at a rate of 5 litres per minute. The mixture is continuously stirred and drained out of the tank at the rate of 10 litres per minute.

   > In this exercise the quantity of solution in the container is not constant.

   a Show that the quantity Q of salt (in grams), t minutes after the start of the experiment, is modelled by $\dfrac{dQ}{dt} + \dfrac{2Q}{20-t} = 50$, $0 \le t < 20$.
   b Hence estimate the quantity of salt in the tank 10 minutes after the start of the experiment.

3.6 Homogeneous differential equations and substitution methods

A homogeneous differential equation is a differential equation that can be written in the form

$$\frac{dy}{dx} = f\left(\frac{y}{x}\right).$$

For example, $\dfrac{dy}{dx} = \dfrac{x^2 - y^2}{xy}$ is a homogeneous first order differential equation because it can be written as $\dfrac{dy}{dx} = \dfrac{1 - \left(\dfrac{y}{x}\right)^2}{\dfrac{y}{x}}$.

You may have guessed the method to solve the equation

$\dfrac{dy}{dx} = \dfrac{1 - \left(\dfrac{y}{x}\right)^2}{\dfrac{y}{x}}$. We substitute $\dfrac{y}{x}$ by a new variable v, i.e. make $y = vx$.

Using the product rule, $\dfrac{dy}{dx} = x\dfrac{dv}{dx} + v$. These homogeneous equations can then be transformed into separable equations of the form $\dfrac{dv}{dx} = \dfrac{f(v) - v}{x}$. We'll now look at an example of this process:

> Homogeneous equations can be easily identified when written in the form $\dfrac{dy}{dx} = f(x, y)$ because all the terms on the RHS must have the same degree. For example, we can see by inspection that $\dfrac{dy}{dx} = \dfrac{x^3 - y^3}{xy^2}$ is homogeneous because all terms on the RHS have degree 3.

Modeling dynamic phenomena

Example 19

Find the general solution of the homogeneous differential equation $\dfrac{dy}{dx} = \dfrac{x^2 - y^2}{xy}$.	

$\dfrac{dy}{dx} = \dfrac{x^2 - y^2}{xy} \Rightarrow \dfrac{dy}{dx} = \dfrac{1 - \left(\dfrac{y}{x}\right)^2}{\dfrac{y}{x}}$	Divide both terms of the RHS by x^2.
$x\dfrac{dv}{dx} + v = \dfrac{1 - v^2}{v} \Rightarrow \dfrac{dv}{dx} = \dfrac{1}{x} \cdot \dfrac{1 - 2v^2}{v}$	Make $y = vx$ and $\dfrac{dy}{dx} = \dfrac{dv}{dx}x + v$ and re-arrange the equation for $\dfrac{dv}{dx}$.
$\dfrac{v}{1 - 2v^2}dv = \dfrac{1}{x}dx \Rightarrow -\dfrac{1}{4}\displaystyle\int \dfrac{-4v}{1 - 2v^2}\,dv = \int \dfrac{1}{x}\,dx$	Separate the variables and integrate both sides.
$-\dfrac{1}{4}\ln\lvert 1 - 2v^2 \rvert = \ln\lvert x \rvert + c$	
$\ln\lvert 1 - 2v^2 \rvert = \ln\lvert x^{-4} \rvert + c \Rightarrow$ $\left\lvert \dfrac{x^2 - 2y^2}{x^2} \right\rvert = A\lvert x^{-4} \rvert$	Rewrite in terms of x and y, using $v = \dfrac{y}{x}$ and simplify.

The example above illustrates the general case, but in some cases the substitution $v = \dfrac{y}{x}$ also transforms the homogeneous equation into a linear equation simply because the classification of first order differential equations are not mutually exclusive. The example below illustrates one of these situations.

Example 20

Show that the substitution $v = \dfrac{y}{x}$ transforms the homogeneous equation $\dfrac{dy}{dx} = \dfrac{2x - y}{x}$ into a first order linear equation and a separable variables differential equation. Hence solve the equation using two different methods.	

$\dfrac{dy}{dx} = \dfrac{2x - y}{x} \Rightarrow \dfrac{dy}{dx} = 2 - \dfrac{y}{x}$	Simplify the RHS
$\dfrac{dv}{dx}x + v = 2 - v$	Make $y = vx$ and $\dfrac{dy}{dx} = \dfrac{dv}{dx}x + v$
$\dfrac{dv}{dx}x + v = 2 - v \Rightarrow \dfrac{dv}{dx} = \dfrac{2 - 2v}{x}$ $\therefore$ separable equation	Re-arrange the equation for $\dfrac{dv}{dx}$ to obtain the form $\dfrac{dv}{dx} = \dfrac{f(x)}{g(v)}$.
$\dfrac{dv}{dx}x + v = 2 - v \Rightarrow \dfrac{dv}{dx} + \dfrac{2}{x}v = \dfrac{2}{x}$ $\therefore$ linear equation	Re-arrange the equation to the form $\dfrac{dv}{dx} + p(x)v = q(x)$.

Method 1	
$\dfrac{dv}{dx} = \dfrac{2-2v}{x} \Rightarrow -\dfrac{1}{2}\int \dfrac{1}{v-1}dv = \int \dfrac{1}{x}dx$	Separate the variables and integrate both sides
$-\dfrac{1}{2}\ln|v-1| = \ln|x| + c \Rightarrow -\dfrac{1}{2}\ln\left|\dfrac{y}{x} - 1\right| = \ln|x| + c$	Rewrite in terms of x and y, using $= \dfrac{y}{x}$, to obtain a solution in implicit form.
$\ln\left|\dfrac{y-x}{x}\right|^{-\frac{1}{2}} = \ln|x| + c \Rightarrow \left|\dfrac{y-x}{x}\right|^{-\frac{1}{2}} = A|x|$	Use properties of logarithms and simplify the equation to obtain y explicitly as a function of x.
$\Rightarrow \dfrac{y-x}{x} = \dfrac{\pm A}{x^2} \Rightarrow y = x + \dfrac{k}{x}$	
Method 2	
$\dfrac{dv}{dx} + \dfrac{2}{x}v = \dfrac{2}{x} \Rightarrow x^2\dfrac{dv}{dx} + 2xv = 2x$	Multiply both sides by the integrating factor $I(x) = e^{\int p(x)dx} = e^{\int \frac{2}{x}dx} = x^2$ to obtain an exact equation.
$\Rightarrow \dfrac{d}{dx}(x^2 v) = 2x$	
$x^2 v = x^2 + c \Rightarrow xy = x^2 + c \Rightarrow y = x + \dfrac{c}{x}$	Integrate both sides, rewrite in terms of x and y (using $= \dfrac{y}{x}$) to obtain a solution in explicit form.

The example above illustrate some important aspects we should consider when solving differential equations:

- For some equations we can use more than one method to solve them;

- The form of the solutions may be very different but we can verify that each of them is indeed a solution either by differentiation or by manipulating one solution to obtain the same form as the solution obtained when the equation is solved using a different method.

Taking the solution we obtained via method 1 in example 20, we can show that

$$-\dfrac{1}{2}\ln\left|\dfrac{y}{x} - 1\right| = \ln|x| + c \Rightarrow \dfrac{dy}{dx} = \dfrac{2x-y}{x}$$

using implicit differentiation.

Taking the solution we obtained via method 2,

$$y = x + \dfrac{c}{x} \Rightarrow \dfrac{dy}{dx} = 1 - \dfrac{c}{x^2}$$ and we can verify

that $\dfrac{dy}{dx} = \dfrac{2x-y}{x}$.

> Substitution techniques are among the most important tools in calculus. When solving differential equations mathematicians often try this approach, although they cannot guarantee in advance that the substitution they try will produce a solution. Trial and error, time, and skill are what is required to find the solution of challenging but important problems. To solve some of these problems, several changes of variables (both independent and dependent) may be needed. If you want to try some of these problems, answer questions 7, 8, and 9 of Exercise 3G.

Exercise 3G

1 Decide whether or not the following equations are homogeneous.

  a $\dfrac{dy}{dx} = \dfrac{3x^2 - xy}{x^2}$

  b $2x\dfrac{dy}{dx} = x - y + 3$

  c $(x^2 + y)\dfrac{dy}{dx} = x$

2 Solve the homogeneous equation in question 1.

3 Use two different methods to solve the following differential equations.

  a $2x\dfrac{dy}{dx} = x - y$
  b $x\dfrac{dy}{dx} = 2x + 3y$

4 Solve the following homogeneous equations.

  a $x^2 \dfrac{dy}{dx} = 4x^2 + y^2$

  b $(x^2 + y^2)\dfrac{dy}{dx} = xy$

  c $x(x^2 + y^2)\dfrac{dy}{dx} = x^3 + 2x^2 y + y^3$

> **EXAM-STYLE QUESTION**
> **5** Find the solution of the following homogeneous equations that satisfy the initial conditions given.
>
>   **a** $(x^2 - y^2)\dfrac{dy}{dx} = xy$ and $y(4) = 2$
>
>   **b** $x^2 \dfrac{dy}{dx} = x^2 - xy + y^2$ and $y(1) = 0$
>
>   **c** $x^2 \dfrac{dy}{dx} = x^2 + xy + y^2$ and $y(1) = 0$

6 The lines of force due to a magnet lying along the y-axis are described by the differential equation $\dfrac{dy}{dx} = \dfrac{2y^2 - x^2}{3xy}$.

  a Use a GDC or computer software to sketch the slope field of this equation.

  b Solve the equation to find the curves on which the iron fillings would align themselves. Add some of the solution curves to your sketch.

EXTENSION QUESTIONS

7 Consider the following differential equation.

$$2xy\frac{dy}{dx} = 1 + y^2 - x^2$$

- **a** Show that this is not homogeneous, nor linear nor separable.
- **b** Show that the substitution $v = y^2$ transforms the equation to

$$x\frac{dv}{dx} = 1 + v - x^2$$

- **c** Show that the substitution $u = x^2$ transforms the equation to

$$2u\frac{dv}{du} = 1 + v - u$$

  Remember: $\frac{dv}{du} = \frac{dv}{dx}\frac{dx}{du}$

- **d** Show that the substitution $w = 1 + v$ transforms the equation to

$$2u\frac{dw}{du} = w - u$$

- **e** Show that the equation in part (d) is both linear and homogeneous.
- **f** Hence solve the equation giving your answer in the form $f(x, y) = 0$. Describe geometrically the curve $f(x, y) = 0$.

8 Consider the following differential equation.

$$\frac{dy}{dx} = \frac{x - y + 2}{x + y}$$

- **a** Transform the equation to a homogeneous one using the following change of variables: $x = u - 1$ and $y = v + 1$.
- **b** Hence solve the differential equation.

9 Differential equations that can be written in the form

$$\frac{dy}{dx} + p(x)y = q(x)y^n$$

where $p(x)$ and $q(x)$ are continuous functions on an interval of real numbers and n is a natural number are part of an important family of equations called **Bernoulli Equations**. You already know how to solve them for $n = 0$ and $n = 1$. For other values of n you can start by dividing both sides of the equation by y^n to obtain an equation of the form $y^{-n}\frac{dy}{dx} + p(x)y^{1-n} = q(x)$ and then use the substitution $v = y^{1-n}$ to obtain a linear differential equation. Use this method to solve the following Bernoulli equations

You may want to research and discover real-life applications of Bernoulli equations.

- **a** $\frac{dy}{dx} + 2y = y^2$
- **b** $\frac{dy}{dx} - y = y^3$

3.7 Euler Method for first order differential equations

In the previous sections we have studied different methods that allowed us to solve many first order differential equations. However, despite the skill of the mathematicians who first discovered these methods, the methods we have looked at so far are not enough to solve all first order differential equations. For example, equations that can be written in the form $\frac{dy}{dx} = f(x, y)$ cannot be solved using the methods we've discussed so far.

You already know how to solve the linear equation $x\frac{dy}{dx} + 3y = e^{x^3}$ (see example 13). However, observe what happens if we try to apply exactly the same method that we used in Example 13 to solve $x\frac{dy}{dx} + 2y = e^{x^3}$:

First re-arrange the equation.

$$x\frac{dy}{dx} + 2y = e^{x^3} \Rightarrow \frac{dy}{dx} + \frac{2}{x}y = \frac{e^{x^3}}{x}$$

Then find the integrating factor.

$$I(x) = e^{\int \frac{2}{x}dx} = e^{2\ln x} = x^2$$

Multiply both sides of the equation by $I(x)$ to obtain an exact equation.

$$x^2 \cdot \left(\frac{dy}{dx} + \frac{2}{x}y\right) = x^2 \cdot \frac{e^{x^3}}{x} \Rightarrow x^2\frac{dy}{dx} + 2xy = x\,e^{x^3}$$

Now, we must integrate both sides.

$$\frac{d}{dx}(x^2 y) = x\,e^{x^3} \Rightarrow x^2 y = \int xe^{x^3}\,dx$$

However, we do not know how to evaluate this integral. You may try different methods to find $\int xe^{x^3}\,dx$, including using a CAS calculator:

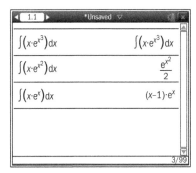

But soon, you will notice that even this calculator cannot produce a general expression for this integral (although it can for other similar integrals).

Chapter 3   85

We have found a limitation of the methods studied: they rely on the possibility of integrating the expressions obtained. This means that we may not be able to find analytic general solutions for all first order differential equations. For equations like $x\frac{dy}{dx} + 2y = e^{x^3}$ we will need to use a different approach: a numerical approach, i.e. produce a table of values at different points that numerically approximate the solution $y = y(x)$.

As numerical methods can be extremely time consuming when done by hand, mathematicians focus on developing algorithms and then use computers to do the hard work.

So let's look at a simpler first order equation, analyse its geometrical meaning, and then re-discover a simple method (Euler's method) to estimate its solution.

Consider the equation $\frac{dy}{dx} = xy$. Geometrically this equation tells us that at any point (x, y) the gradient or slope of the tangent to the solution $y = y(x)$ has the value xy. We can visualize this by plotting the slope field:

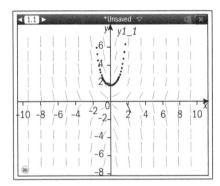

If we enter an initial value, let's say $y(0) = 1$, we can also visualise a particular solution. With an interactive GDC we can drag the initial point and explore other particular solutions for different initial conditions.

This quick calculation produced by the calculator within seconds can be done slowly by hand using the following **algorithm**:

Given a first order differential equation written in the form
$\frac{dy}{dx} = f(x, y)$,

Step 1: Select the starting point (x_0, y_0).
Step 2: Select the x-coordinate x_n of the end point and the number N of intermediate points you want to calculate. Let $h = \frac{x_n - x_0}{N}$.
Step 3: Use the recurrence formulas
$x_{n+1} = x_n + h$ and $y_{n+1} = y_n + hf(x_n, y_n)$ for $n = 0, 1, ..., N-1$
Step 4: Plot the points $(x_0, y_0), (x_1, y_1), ..., (x_N, y_N)$

Modeling dynamic phenomena

This algorithm is called **Euler's Method** or **Method of the Tangents** as the value of the successive y-coordinates are estimated assuming that locally the solution behaves like its tangent at (x_n, y_n) and the value of y_{n+1} is obtained by moving h units along this tangent to the next point (x_{n+1}, y_{n+1}).

Example 21

Apply Euler's method with step size $h = 0.25$ to approximate the solution to the initial value problem $\frac{dy}{dx} = xy$ and $y(0) = 1$. Find the coordinates of five points and sketch the graph of an approximate solution $y = y(x)$.

n	x_n	y_n	$f(x_n, y_n)$
0	0	1	0
1	0.25	1	0.25
2	0.5	1.0625	0.53125
3	0.75	1.1953125	0.896484375
4	1	1.419433594	1.419433594

Make $x_0 = 0$ and $y_0 = 1$.

Use the recurrence formulas
$x_{n+1} = x_n + h$ and
$y_{n+1} = y_n + hf(x_n, y_n)$
for $n = 0, 1, 2, 3, 4$

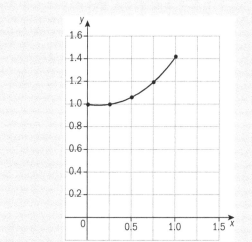

Plot the points (x_n, y_n) and connect them with a smooth line.

Let's now solve numerically the differential equation $x\frac{dy}{dx} + 2y = e^{x^3}$ using Euler's method. The following example shows you how to use a GDC to obtain both the solution to an initial value problem, and the graph of this solution.

Example 22

Apply Euler's method with step size $h = 0.2$ to approximate the solution to the initial value problem $x\dfrac{dy}{dx} + 2y = e^{x^3}$ and $y(-2) = 1$. Use technology to obtain the graph of the solution of this initial value problem and estimate the value of $y(-1)$.

$x\dfrac{dy}{dx} + 2y = e^{x^3} \Rightarrow \dfrac{dy}{dx} = \dfrac{e^{x^3} - 2y}{x}$	Write the differential equation in the form $\dfrac{dy}{dx} = f(x, y)$.

Use technology to apply the algorithm:

Make $x_0 = -2$ and $y_0 = 1$ and set the recurrence formulas

$$x_{n+1} = x_n + h \text{ and}$$
$$y_{n+1} = y_n + hf(x_n, y_n)$$
for $n = 0, 1, 2, 3, 4$

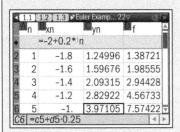 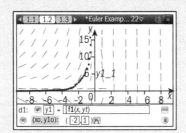 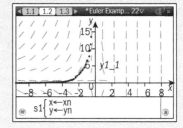

∴ $y(-1) = 3.97$ according to Euler's method with step 0.2; comparing the graph obtained from the GDC with the scatter plot of the points (x_n, y_n), this approximation looks good.

Investigation: Approximate numerical solutions versus exact solutions

Use your GDC or a spreadsheet to perform the calculations required below.

Consider the equation $\dfrac{dy}{dx} = xy$. Apply Euler's method to determine $y(1)$ starting at $y(0) = 1$ using:

a 5 iterations (N = 5)
b 10 iterations (N = 10)
c 20 iterations (N = 20)

Plot the points found and sketch the graph of an approximate solution $y = y(x)$ in each case.

d Solve the separable equation $\dfrac{dy}{dx} = xy$ and draw the graph of the analytic solution on the same axes. You should use the initial condition $y(0) = 1$. Comment on the results.

e Repeat parts (a)-(c) starting at $y(1) = 2$. Compare the graph obtained with the analytical solution of $\dfrac{dy}{dx} = xy$ when $y(1) = 2$. Comment of the results in relation to the ones obtained in part (d).

As we have seen from the investigation, the solutions obtained using Euler's Method are less accurate than the true solutions. This can cause problems! There are two important factors to monitor: firstly, the effect of rounding the intermediate values, whose effect on the final value increases as we decrease the value of the step h (because we introduce more rounded values); and secondly, the effect of the truncation or linear approximation generated by the method itself, that can be minimized by decreasing the value of h. Since increasing or decreasing the step h has an opposite effect on the two types of error, it is difficult to predict the optimal value for h. In chapter 5, you will extend your knowledge of approximation methods and be able to compute more accurate approximations of values of a function using its derivatives.

Short Investigation - approximation of Euler's number

Solve the equation $\frac{dy}{dx} = y$ when $y(0) = 1$. Hence explain how you can use Euler's Method to obtain an approximation of e. Use smaller and smaller values of h and comment on the results obtained. Draw diagrams to explain your reasoning.

> In the Winter of 1961 at MIT, Professor Lorenz was running a climate model consisting of twelve differential equations representing several parameters, when he decided to re-examine one of the model's sequences. From a printout, he took conditions from a mid-point in the model run and reinitiated the calculations, making only one slight change: the original inputs had six decimal places, but in the second run Lorenz, to save time and space, rounded them to three decimal places. Lorenz expected that his second run would match the first, but it didn't. It was, in fact, almost precisely the same at the beginning, but then the second run diverged radically from the first.
>
> Lorenz first suspected a hardware problem, but then realized the truth: the rounding of the initial inputs – a tiny change in initial values – had produced wildly divergent results. Lorenz realized that it is not possible to predict weather in the long term because of the climate's 'sensitive dependence on initial conditions'. He described it as 'The Butterfly Effect' – a perfect choice of terms given the graphic which the Lorenz Strange Attractor, with its fractal dimension, generates.
>
>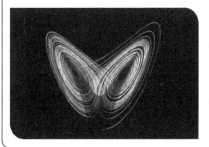
>
> The Lorenz Attractor, as part of Chaos Theory, is helping medicine to better understand heart attacks, and predict them before they occur.

Exercise 3H

1. Apply Euler's method with step size $h = 0.1$ to approximate the solution to the initial value problem $\frac{dy}{dx} = y$ and $y(0) = 1$. Use technology to obtain the slope field and the graph of the solution of this initial value problem. Estimate the value of $y(1)$.

2. Consider the differential equation $\frac{dy}{dx} = 2x + y$.
   a. Draw the isoclines $2x + y = m$ for $m = 0, \pm 1, \pm 2, \pm 3$.
   b. Hence draw the slope field for this differential equation.
   c. Use Euler's Method to approximate the value of $y(1)$ when $y(0) = 1$
   d. Plot the points (x_n, y_n) obtained with the Euler method in part (b).
   e. Solve the differential equation $\frac{dy}{dx} = 2x + y$ and find the exact value of $y(1)$.
   f. Compare the values of $y(1)$ found in parts (c) and (f).

3. Consider the differential equation $\frac{dy}{dx} = x^2 - y^2$.
   a. Use Euler's Method to approximate the value of $y(-1)$ when $y(-2) = 1$.
   b. Use technology to plot the slope field and graph the solution found in part (a).

4. a. Use Euler's method to find an approximate solution to $\frac{dy}{dx} = \frac{x}{\sin(y)e^x}$ with initial condition $(2, 1)$ at the point $x = 3$. Use $h = 0.1$.
   b. Compare your answer with the exact value obtained when the equation is solved (you may use example 11 for a general solution to the equation).
   c. A computer algebra system was used to obtain the solution to this equation graphically. Observe the graph and comment on the limitations of Euler's Method.

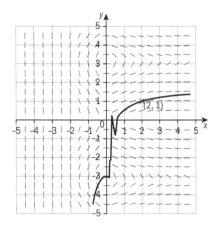

5 Consider the differential equation $\dfrac{dy}{dx} = \dfrac{-2xy}{1+x^2}$, where $y = 1$ when $x = 2$.

   a Use Euler's Method with $h = 0.1$ to find an approximate solution of y when $x = 3$, giving your answer to two decimal places.

   b **Solve** the differential equation $\dfrac{dy}{dx} = \dfrac{-2xy}{1+x^2}$ and find the solution $y = f(x)$ that contains the point $(2, 1)$.

   c **Find** $f(3)$ and compare your answer with the value found in (a).

6 a Using Euler's Method, find an approximate value of the solution to the equation $y' = y^2$ at $x = 0.25, 0.5, 0.75,$ and 1, given that $y(0) = 1$.

   b Solve the differential equation and find the exact values of $y(x)$ for $x = 0.25, 0.5,$ and 0.75.

   c Hence comment on the values of the errors obtained.

7 Stefan's law of cooling states that $\dfrac{dT}{dt} = -k(T^4 - M^4)$ where T is the temperature of a body; M is the temperature of the medium where the body is; t the time in hours; and k a constant. Temperatures are measured in °C.

   a Assuming that $k = 5 \times 10^{-6}$, explain the meaning of the differential equation.

   b Given that a room has constant temperature M, where $M = 20$ and $T(0) = 35$, use Euler's Method and a step $h = 0.25$ to approximate $T(1)$.

   c Show that
   $$\ln\dfrac{T+M}{T-M} + 2\arctan\dfrac{T}{M} - 4M^3 kt = c$$
   is solution of the differential equation.

   Hence find the equation of the curve that contains the point $(0, 35)$ when $k = 0.05$ and $M = 20$ and calculate the exact value of $T(1)$.

   d State how to proceed to minimize the approximation error when using Euler's Method.

8 Consider the differential equation $\dfrac{dy}{dx} = e^{-xy}$.

   a Use Euler's Method with step $h = 0.25$ to find five points on the curve of the solution to the differential equation, starting at the origin.

   b Use a GDC or computer software to sketch the slope field for the differential equation.

Review exercise

EXAM-STYLE QUESTIONS

1. The temperature $T\,°C$ of an object in a room, after t minutes, satisfies the differential equation
   $$\frac{dT}{dt} = k(T - 23),\text{ where } k \text{ is a constant.}$$

   a Solve this equation to show that $T = Ae^{kt} + 23$, where A is a constant.

   b When $t = 0$, $T = 90$, and when $t = 10$, $T = 70$.
      i Use this information to find the value of A and of k.
      ii Hence find the value of t when $T = 40$.

2. A sample of radioactive material decays at a rate which is proportional to the amount of material present in the sample. Find the half-life of the material if 50 grams decay to 47 grams in 10 years.

3. Solve the differential equation $x\dfrac{dy}{dx} - y^2 = 1$, given that $y = 0$ when $x = 3$. Give your answer in the form $y = f(x)$, for some function $f(x)$.

4. The function $y = f(x)$ satisfies the differential equation
   $$2x^2 \frac{dy}{dx} = x^2 + y^2 \ (x > 0)$$

   a Using the substitution $y = vx$, show that $v + x\dfrac{dv}{dx} = \dfrac{1+v^2}{2}$.

   b Hence show that this is an equation with separable variables.

   c Find the general solution of the original differential equation.

5. Find the solution to the differential equation $\dfrac{dy}{dx} = 1 + 2y\tan(x)$ that satisfies the initial condition $y\left(\dfrac{\pi}{4}\right) = 2$.

6. Consider the differential equation $\dfrac{dy}{dx} = \dfrac{3y^2 + x^2}{2xy}$, for $x > 0$.

   a Use the substitution $y = vx$ to show that $x\dfrac{dv}{dx} = \dfrac{1+v^2}{2v}$.

   b Hence find the general solution of the differential equation.

   c Find an explicit expression $y = f(x)$ for the solution to the equation that satisfies the boundary conditions $y = 2$ when $x = 1$, stating clearly its largest possible domain.

   d Sketch the graph $y = f(x)$, showing clearly that this solution satisfies the boundary conditions.

7 Show that the general solution to the differential equation
$(x^2 + y^2) + 2xy \dfrac{dy}{dx} = 0$, $x > 0$ is $x^3 + 3xy^2 = a$, where a is a constant.

8 Consider the first order linear differential equation
$$\dfrac{dy}{dx} = y \tan x + 1, \ 0 < x < \dfrac{\pi}{2}.$$
  a Show that an integrating factor for this equation is $I(x) = \cos(x)$.
  b Hence solve the equation and find the solution that satisfies $y(0) = 2$.

9 Consider the differential equation $\dfrac{dy}{dx} + \dfrac{xy}{4 - x^2} = 2$, where $-2 < x < 2$ and $y(0) = 1$.
  a Use Euler's Method with $h = 0.2$ to find an approximate value of y when $x = 0.6$, giving your answer to three decimal places.
  b By first finding an integrating factor, solve this differential equation. Give your answer in the form $y = f(x)$.
  c Calculate, correct to three decimal places, the value of y when $x = 0.6$.
  d Sketch the graph of $y = f(x)$ for $0 < x < 0.6$. Use your sketch to explain why your approximate value of y is greater than the true value of y.

10 When air is released from an inflated balloon it is found that the rate of decrease of the volume of the balloon is proportional to the volume of the balloon. This can be modelled by the differential equation $\dfrac{dV}{dt} = -\alpha V$ where V is the volume of the balloon, t is the time, and α is the constant of proportionality.
  a If the initial volume of the balloon is V_0, find an expression, in terms of α, for the volume of the balloon at time t.
  b Find an expression, in terms of α, for the time when the volume is $\dfrac{V_0}{3}$.

11 Find the solution $y = f(x)$ to the differential equation $xy \dfrac{dy}{dx} = 1 + y^2$ that satisfies the initial conditions $y = 0$ when $x = 2$. Hence sketch the graph of $y = f(x)$ and state clearly its largest possible domain.

EXAM-STYLE QUESTIONS

12 The equation of motion of a particle with mass m, subjected to a force kx, can be written as $kx = mv\dfrac{dv}{dx}$, where x is the displacement and v is the velocity. Given that $v(0) = v_0$, Find v (in terms of v_0), k, and m, when $x = 1$.

13 Find the values of a and b that satisfy the identity $\dfrac{1}{4x - x^2} = \dfrac{a}{x} + \dfrac{b}{x - 4}$

Hence solve the differential equation $\dfrac{dx}{dt} = kx(4 - x)$ where $0 < x < 4$, and k is a constant.

14 A uniform cable of length l metres is placed with its ends on two supports A and B, which are at the same horizontal level.

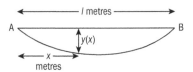

If $y(x)$ metres is the distance below [AB] at which the cable hangs when it is at a distance x metres from support A, then it is known that $\dfrac{d^2y}{dx^2} = \dfrac{1}{125l^3}(x^2 - lx)$.

a Let $z = \dfrac{1}{125l^3}\left(\dfrac{x^3}{3} - \dfrac{lx^2}{2}\right) + \dfrac{1}{1500}$. Show that $\dfrac{dz}{dx} = \dfrac{1}{125l^3}(x^2 - lx)$.

b Given that $\dfrac{dy}{dx} = z$ and $y(0) = 0$, find an expression for $y(x)$.

c Hence show that y satisfies $\dfrac{d^2y}{dx^2} = \dfrac{1}{125l^3}(x^2 - lx)$.

d Show that $y(l) = 0$.

e Given that $l = 4$, find the value of $y(2)$ and state its meaning in the context of the problem.

15 Consider the differential equation $\dfrac{dy}{dx} + \dfrac{2y}{x\ln(x)} = \dfrac{1 - x}{(\ln(x))^2}$, $x \in \mathbb{R}^+ \setminus \{1\}$.

a Find the general solution of the differential equation.

b Hence, find the particular solution $y = f(x)$ that contains the point (e, e).

c Sketch the graph of your solution for $x > 1$, clearly indicating any asymptotes and axes intercepts.

Chapter 3 summary

Differential equations are a type of a class of equations where the **unknowns are functions.** A differential equation describes a **relation between the unknown function, its derivatives, the independent variable and some constants.** The highest derivative order that appears on a differential equation is called **the order of the equation.** Differential equations can also be classified as **linear or non-linear.** Non-linear equations include expressions like y^2, $(y')^2$, $\sin(xy)$, e^y, …

Solving the three types of first order differential equations:

1 **First order differential equations with separable variables** can be written in the form $\frac{dy}{dx} = \frac{f(x)}{g(y)}$. To solve, re-arrange as $g(y)\frac{dy}{dx} = f(x)$ and then integrate both sides to obtain an implicit general solution of the equation $\int g(y)\,dy = \int f(x)\,dx + C$.

2 **First order linear equations** can be written in the form $\frac{dy}{dx} + p(x)y = q(x)$.

   To solve, multiply both sides by the integrating factor $I(x) = e^{\int p(x)dx}$ to obtain an exact equation of the form $\frac{d}{dx}(u(x) \cdot y) = v(x)$ and then integrate to obtain a general solution.

3 **First order homogeneous differential equations** can be written in the form $\frac{dy}{dx} = f\left(\frac{y}{x}\right)$. Use the substitution $y = vx$ to transform the equation into separable variables of the form $\frac{dv}{dx} = \frac{f(v) - v}{x}$. Sometimes this substitution also transforms homogeneous equations into linear equations.

It some cases it is possible to find **explicit solutions** to differential equation and, if a boundary condition (initial condition of the form $y(x_0) = y_0$) is given, determine the particular solution $y = y(x)$ whose graph contains the point (x_0, y_0). In most cases however it just possible to find implicit solutions of the differential equation.

Euler's method or method of the tangents – apply the following algorithm:

Step 1: Select the starting point (x_0, y_0).
Step 2: Select the x-coordinate x_n of the end point and the number N of intermediate points you want to calculate. Let $h = \frac{x_n - x_0}{N}$.
Step 3: Use the recurrence formulas.
$\quad\quad x_{n+1} = x_n + h$ and $y_{n+1} = y_n + hf(x_n, y_n)$ for $n = 0, 1, …, N-1$
Step 4: Plot the points $(x_0, y_0), (x_1, y_1), …, (x_N, y_N)$

A **slope field** or **phase graph** is a diagram that represents a differential equation $\frac{dy}{dx} = f(x, y)$ and shows at each point (x, y) a segment representing the slope (or gradient) of the tangent to the graph of the solutions of the differential equation. Slope fields allow prediction and graphical approximations of solutions to differential equation.

The **term isocline** derives from the Greek words for 'same slope'. For a first-order differential equation $\frac{dy}{dx} = f(x, y)$, a curve with equation $f(x, y) = m$, for some constant m, is known as an isocline. All the solutions of the ordinary differential equation intersecting that curve have the same slope m. Isoclines can be used as a graphical method of solving differential equations.

The finite in the infinite

CHAPTER OBJECTIVES:

9.2 Convergence of infinite series; tests for convergence: comparison test, limit comparison test, ratio test, integral test; the *p*-series; series that converge absolutely; series that converge conditionally; alternating series; power series.

9.4 The integral as a limit of a sum; lower and upper Riemann sums; fundamental theorem of calculus; improper integrals.

Before you start

You should know how to:

1 Find the limit of $\{u_n\}$ as $n \to \infty$

e.g. find $\lim\limits_{n\to\infty} \dfrac{2n^2+3n}{3n^2-2n+1}$.

$$\lim_{n\to\infty} \frac{2n^2+3n}{3n^2-2n+1} = \lim_{n\to\infty} \frac{2+\dfrac{3}{n}}{3-\dfrac{2}{n}+\dfrac{1}{n^2}} = \frac{2}{3}$$

2 Determine if sequences converge by attempting to find the limits, e.g. $u_n = \dfrac{1}{n!}$.

Applying the squeeze theorem,

$0 \le \dfrac{1}{n!} \le \dfrac{1}{2^n}$, for all $n \ge 4$.

Since $\lim\limits_{n\to\infty} \dfrac{1}{2^n} = 0$, then $\lim\limits_{n\to\infty} \dfrac{1}{n!} = 0$.

3 Use all core syllabus integration techniques to find indefinite and definite integrals,

e.g. find $\displaystyle\int \dfrac{1}{x \ln x}\,dx$.

Using substitution, let $u = \ln x$, then $\dfrac{du}{dx} = \dfrac{1}{x}$.

Hence $\displaystyle\int \dfrac{1}{x \ln x}\,dx = \int \dfrac{1}{u}\,du = \ln u = \ln(\ln x) + c$.

Skills check:

1 Find the limits, if they exist:

a $\lim\limits_{n\to\infty} \dfrac{n}{n+1}$ **b** $\lim\limits_{n\to\infty} \dfrac{-3n^3+2n}{1-3n-4n^3}$

c $\lim\limits_{n\to\infty} \dfrac{5n^2}{3n-1}$ **d** $\lim\limits_{n\to\infty} \dfrac{e^n-1}{e^n+1}$

2 Determine if the sequence converges.

a $u_n = \dfrac{\sin n}{n}, n > 0$

b $u_n = \dfrac{n!}{n}, n > 0$

c $u_n = (\sqrt{n+2} - \sqrt{n}), n \ge 0$

3 Integrate

a $\displaystyle\int (\sqrt{x+2} - \sqrt{x})\,dx$

b $\displaystyle\int \dfrac{e^x}{1+e^{2x}}\,dx$

c $\displaystyle\int \dfrac{\sin x}{x}\,dx$

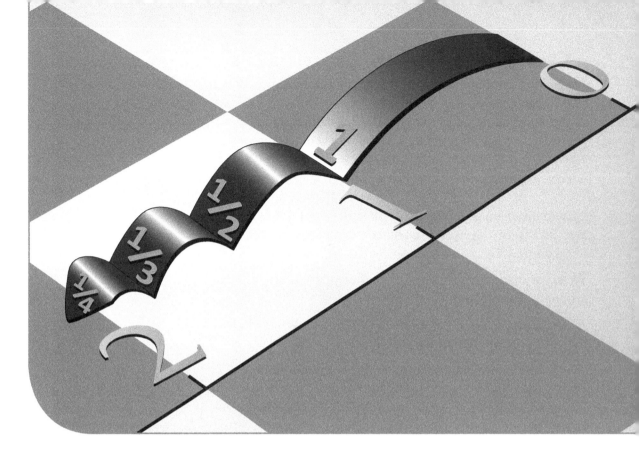

Convergence of infinite series

"The infinite! No other question has ever moved so profoundly the spirit of man." —**David Hilbert, German Mathematician, 1862–1943**

The question as to how an infinite sum of positive terms can yield a finite result (such as the geometric series that you have studied) posed a deep philosophical and mathematical challenge. It was an important gap to close in understanding the mathematics of infinity. Since infinite series were used throughout the development of calculus, it is difficult to pinpoint their exact historical path.

While infinite series began to be used as early as the 14th century, mathematicians dealt mainly with intuitive notions of convergence and divergence. Intuition, however, proved to be most unreliable in determining convergence. Nonetheless, mathematicians such as Newton, Taylor, Leibniz, and Euler appeared unconcerned by this. The French mathematician Cauchy (whose name you have already seen in the first few chapters) was the first to give a rigorous definition of convergence of a series by establishing his "Cauchy Criterion". His work was truly revolutionary, and made contemporary mathematicians nervous about "getting it right" when working with series. It is said that the French mathematician and physicist Laplace went into seclusion before publishing his famous work *Celestial Mechanics* in order to check if all the series in his book lived up to Cauchy's scrutiny. Fortunately, they did!

> The beginnings of the concept of infinite sequences and series can be found in works of mathematicians of the Kerala School in southern India as early as the 14th century. It is claimed that Isaac Newton's idea of calculus came from the Kerala School, after it was stolen in 1602 and transported back to Europe by the East India Company's Dutch wing.

4.1 Series and convergence

When an infinite series has a real number as a sum, we say that it is a **convergent series**, and it converges to its sum. You have met convergent infinite series before in the core syllabus (e.g. the geometric series). We know that an infinite geometric series will converge if its common ratio is between −1 and 1, i.e. $|r|<1$, in which case its sum is $\frac{u_1}{1-r}$, where u_1 is the first term of the series.

Let us now examine more closely the series $1 - 1 + 1 - 1 + 1 - \ldots$, and discuss its convergence. In other words, does this series have a unique sum?

We know that we can arrive at two different sums by a different grouping. For example, $(1 - 1) + (1 - 1) + = \ldots = 0$.

Alternatively, $1 + (-1 + 1) + (-1 + 1) + \ldots = 1$. We conclude therefore that this series diverges.

The Swiss mathematician Leonhard Euler examined this series from a different perspective. Similar to the way we obtained the formula for the sum of a geometric series, he started with the following.

Let $S = 1 - 1 + 1 - 1 + \ldots$

Then, $1 - S = 1 - (1 - 1 + 1 - 1 + \ldots) = 1 - 1 + 1 - 1 \ldots = S$

Hence, since $1 - S = S$, then $S = \frac{1}{2}$.

> Such a series is called a Cesàro series after the Italian mathematician **Ernesto Cesàro (1859–1906).** The series may or may not converge; but if it does converge, then its sum is referred to as a Cesàro sum.

To investigate this series further, we can consider the sequence of partial sums.

Definition: The **partial sums** of a series are defined as

$S_1 = u_1;\ S_2 = u_1 + u_2;\ S_3 = u_1 + u_2 + u_3;\ S_n = u_1 + u_2 + u_3 + \ldots + u_n.$

These partial sums form a sequence of partial sums, $S_n = \sum_{k=1}^{n} u_k$.

Hence, $S_1 = 1;\ S_2 = 1 - 1 = 0;\ S_3 = 1 - 1 + 1 = 1;\ S_4 = 0;$ and so on.

The sequence of partial sums $S = 0, 1, 0, 1, \ldots$ oscillates between 0 and 1, and therefore the sequence has no limit.

We can now state what it means for a series to converge.

Definition: If the sequence of partial sums has a limit L as $n \to \infty$, then the series **converges** to the limit L,

i.e. $u_1 + u_2 + u_3 + \ldots + u_n + \ldots = \sum_{k=1}^{\infty} u_k = L.$

Otherwise the series **diverges**.

Since the sequence of partial sums of $1 - 1 + 1 - 1 + \ldots$ oscillate between 0 and 1, the sequence does not have a limit, and hence the series diverges.

Consider the series $1 - 2 + 3 - 4 + \ldots$ Does this series converge? Before considering the partial sums, when attempting a method of shifting and term-by-term addition, we actually obtain that $S = \frac{1}{4}$. (You may want to try doing this on your own!)

However, to answer the question of convergence, we must determine if the sequence of partial sums has a limit as $n \to \infty$.

$S_1 = 1$; $S_2 = 1 - 2 = -1$; $S_3 = 1 - 2 + 3 = 2$; $S_4 = 1 - 2 + 3 - 4 = -2$; and so on. Clearly, the sequence of partial sums $1, -1, 2, -2, \ldots$ does not have a limit, hence the series diverges.

Using this condition to test convergence of series, we can now prove that an infinite geometric series is convergent if $|r| < 1$.

Theorem 1: If $|r| < 1$, then the infinite geometric series
$1 + r + r^2 + r^3 + \ldots + r^n + \ldots$ converges, and
$1 + r + r^2 + r^3 + \ldots + r^n + \ldots = \frac{1}{1-r}$.

Proof: For each non-negative n, $(1-r)(1+r+r^2+\ldots+r^n) = 1 - r^{n+1}$ by expanding and simplifying.

The nth partial sum, S_n, is $1 + r + r^2 + r^3 + \ldots + r^n = \frac{1-r^{n+1}}{1-r}$. We now take the limit of this expression as $n \to \infty$. Since $|r| < 1$, $r^n \to 0$ and we obtain $\frac{1}{1-r}$.

Example 1

Verify that the series $2 + \frac{3}{2} + \frac{9}{8} + \frac{27}{32} + \ldots$ converges, and find its sum.	
This is a geometric series with $r = \frac{3}{4}$, hence it converges.	Definition of geometric series, and it converges since $|r| < 1$
$S_\infty = \dfrac{2}{1-\frac{3}{4}} = 8$	It converges to its sum, $S_\infty = \dfrac{u_1}{1-r}$.

As we saw in the proof of Theorem 1, finding S_∞ is identical to obtaining an expression for the nth partial sum of this series and finding its limit as $n \to \infty$, i.e. $S_n = \dfrac{2\left(1-\left(\frac{3}{4}\right)^n\right)}{1-\frac{3}{4}} = 8\left(1-\left(\frac{3}{4}\right)^n\right)$, and

$\lim\limits_{n \to \infty} S_n = \lim\limits_{n \to \infty} 8\left(1-\left(\frac{3}{4}\right)^n\right) = 8$.

We can see intuitively that if the limit as $n \to \infty$ of the sequence of partial sums exists, then the greater the number of terms, the smaller the terms must be, i.e. $\lim_{n\to\infty} u_n = 0$.

In other words, if for any sequence $u_1 + u_2 + u_3 + \ldots + u_n + \ldots = S$, then the limit of the partial sums exists, i.e. $\lim_{n\to\infty} S_n = S$. In the same manner, $\lim_{n\to\infty} S_{n-1} = S$. Subtracting these two, $\lim_{n\to\infty} S_n - \lim_{n\to\infty} S_{n-1} = \lim_{n\to\infty}(S_n - S_{n-1}) = 0$. Since $u_n = S_n - S_{n-1}$ it follows that $\lim_{n\to\infty} u_n = 0$.

This condition is a **necessary** one for the convergence of infinite geometric series, but it is not a **sufficient** condition. If this condition holds, the series may or may not converge. However, if we know that the series converges, then the limit of the nth term of its sequence must be 0. The contrapositive statement is also true – if the limit of the nth term of the sequence is not 0, then the series does not converge. In determining the convergence of a series, the first step is to see if it meets the necessary condition, i.e. the limit of the nth term of the sequence is 0.

> The **contrapositive** of the statement $p \Rightarrow q$ is $\neg q \Rightarrow \neg p$.

The nth term test for divergence: If the sequence $\{u_n\}$ does not converge to 0, then the infinite series $\sum u_n$ diverges.

> This test is also known as the Divergence Test, or the Vanishing Condition.

Example 2

For the following series, determine if possible whether the series converges or diverges.

a $\dfrac{1}{2} + \dfrac{2}{3} + \dfrac{3}{4} + \ldots + \dfrac{n}{n+1} + \ldots$ **b** $1 + \dfrac{1}{2} + \dfrac{1}{3} + \ldots + \dfrac{1}{n} + \ldots$

---

a $\lim_{n\to\infty} u_n = \lim_{n\to\infty} \dfrac{n}{n+1} = \lim_{n\to\infty} \dfrac{1}{1+\dfrac{1}{n}} = 1$

$\lim_{n\to\infty} \dfrac{n}{n+1} \neq 0$, hence the series diverges.

Using the algebraic method to evaluate the limit of the nth term, i.e. divide numerator and denominator by the largest power of n.

Since the limit of the nth term of the sequence is not 0, the necessary condition for convergence of the series is not met.

b $\lim_{n\to\infty} \dfrac{1}{n} = 0$, hence we cannot say whether the series converges or diverges

The necessary condition for convergence is met, but it is not enough to determine if the series converges.

The finite in the infinite

The series in part **b** of Example 2 is referred to as the **harmonic series**, and it does in fact diverge. The graph below shows that the sequence of partial sums is slowly increasing.

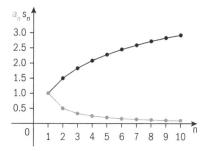

Later in the chapter, we will use calculus techniques to prove that the harmonic series diverges, but we can also prove that it diverges algebraically, using proof by contradiction. A version of this proof was first published in around 1350 by the French monk Oresme.

Assume that $\sum_{n=1}^{\infty} \frac{1}{n} = S$. Then

$$S = 1 + \frac{1}{2} + \frac{1}{3} + \frac{1}{4} + \frac{1}{5} + \frac{1}{6} + \frac{1}{7} + \frac{1}{8} + \frac{1}{9} + \frac{1}{10} + \ldots + \frac{1}{n} + \ldots$$

$$= \left(1 + \frac{1}{2}\right) + \left(\frac{1}{3} + \frac{1}{4}\right) + \left(\frac{1}{5} + \frac{1}{6}\right) + \left(\frac{1}{7} + \frac{1}{8}\right) + \left(\frac{1}{9} + \frac{1}{10}\right) + \ldots + \left(\frac{1}{n} + \frac{1}{n+1}\right) + \ldots$$

$$> \left(\frac{1}{2} + \frac{1}{2}\right) + \left(\frac{1}{4} + \frac{1}{4}\right) + \left(\frac{1}{6} + \frac{1}{6}\right) + \left(\frac{1}{8} + \frac{1}{8}\right) + \left(\frac{1}{10} + \frac{1}{10}\right) + \ldots + \left(\frac{1}{2n} + \frac{1}{2n}\right) + \ldots$$

$$= 1 + \frac{1}{2} + \frac{1}{3} + \frac{1}{4} + \frac{1}{5} + \ldots + \frac{1}{n} + \ldots = S$$

This means that $S > S$, an absurd conclusion! Hence our assumption is false, and the harmonic series does not have a finite sum.

> For an interesting discussion on why this series is called harmonic, as well as its many occurrences in real life situations, see:
> **http://www.plus.maths.org/content/perfect-harmony**

> ❓ We now know that the series of the reciprocals of natural numbers diverges. We will prove later in the chapter, using the convergence tests, that the series of the reciprocal of the squares of natural numbers, $\sum_{n=1}^{\infty} \frac{1}{n^2}$, converges. Finding its sum proved to be extremely difficult. Jakob Bernoulli (of the famous Bernoulli family of mathematicians) attempted to find its sum, and failed. Mengoli and Leibniz also made attempts that failed. The quest for its sum became known as the **Basel Problem**, and was finally found successfully by Euler using power series.

We will now explore the convergence of $\sum_{n=1}^{\infty} \dfrac{2}{4n^2 - 1}$. Looking at the sequence of partial sums on the GDC, we obtain

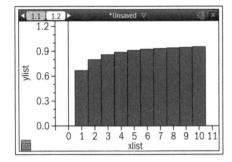

From these GDC screenshots, we can conjecture that this series converges to 1, but how can we be sure that it does?

To check, we would perform partial fraction decomposition, which is not in the syllabus. If this were an exam question, the decomposition would have to be given. So, the question might be worded as follows:

'Show that $\dfrac{2}{4n^2 - 1} = \dfrac{1}{2n-1} - \dfrac{1}{2n+1}$, and hence determine if $\sum_{n=1}^{\infty} \dfrac{2}{4n^2 - 1}$ converges. If it converges, find its sum.'

The first part is easy to show algebraically.

Now, writing out the first few terms, we obtain

$$S_n = \left(1 - \frac{1}{3}\right) + \left(\frac{1}{3} - \frac{1}{5}\right) + \left(\frac{1}{5} - \frac{1}{7}\right) + \ldots + \left(\frac{1}{2n-1} - \frac{1}{2n+1}\right)$$

$$= 1 - \left(\frac{1}{3} - \frac{1}{3}\right) - \left(\frac{1}{5} - \frac{1}{5}\right) - \left(\frac{1}{7} - \frac{1}{7}\right) - \ldots - \left(\frac{1}{2n-1} - \frac{1}{2n-1}\right) - \frac{1}{2n+1}$$

$$= 1 - \frac{1}{2n+1}.$$

Now, $\lim_{n \to \infty}\left(1 - \dfrac{1}{2n+1}\right) = 1$.

Hence, the series converges, and its sum is 1.

The original series above is called a telescoping series, since after converting it into the sum (or difference) of two fractions, and after some cancellations, only the first and last terms remain.

Example 3

Show that $\dfrac{1}{n^2+4n+3}$ can be written as $\left(\dfrac{1}{2n+2}-\dfrac{1}{2n+6}\right)$, and hence determine whether or not $\displaystyle\sum_{n=1}^{\infty}\dfrac{1}{n^2+4n+3}$ converges. If the series converges, find its sum.

$\left(\dfrac{1}{2n+2}-\dfrac{1}{2n+6}\right)=\dfrac{(2n+6)-(2n+2)}{(2n+2)(2n+6)}=\dfrac{4}{4n^2+16n+12}=\dfrac{4}{4(n^2+4n+3)}=\dfrac{1}{n^2+4n+3}$	Fraction addition
$\left(\dfrac{1}{2n+2}-\dfrac{1}{2n+6}\right)=\dfrac{1}{2}\left(\dfrac{1}{n+1}-\dfrac{1}{n+3}\right)$	Factor out $\dfrac{1}{2}$, and find the first few terms of S_n
$S_n=\dfrac{1}{2}\left[\left(\dfrac{1}{2}-\dfrac{1}{4}\right)+\left(\dfrac{1}{3}-\dfrac{1}{5}\right)+\left(\dfrac{1}{4}-\dfrac{1}{6}\right)+\ldots+\left(\dfrac{1}{n}-\dfrac{1}{n+2}\right)+\left(\dfrac{1}{n+1}-\dfrac{1}{n+3}\right)+\ldots\right]$	
$=\dfrac{1}{2}\left[\dfrac{1}{2}+\dfrac{1}{3}-\dfrac{1}{n+2}-\dfrac{1}{n+3}\right]$	
$\displaystyle\lim_{n\to\infty}\left[\dfrac{1}{2}\left(\left(\dfrac{5}{6}\right)-\dfrac{1}{n+2}-\dfrac{1}{n+3}\right)\right]=\dfrac{5}{12}$. Hence, the series converges and its sum is $\dfrac{5}{12}$.	Find the limit of S_n as n approaches infinity.

Exercise 4A

1 Determine, if possible, whether or not the following series converge or diverge. If the series converges, find its sum.

a $\displaystyle\sum_{n=0}^{\infty}3^n$ **b** $\displaystyle\sum_{n=0}^{\infty}\dfrac{1}{3^n}$ **c** $\displaystyle\sum_{n=0}^{\infty}\dfrac{2n!}{3n!+3}$ **d** $\displaystyle\sum_{n=0}^{\infty}3\left(\dfrac{1}{7}\right)^{n-1}$

e $\displaystyle\sum_{n=1}^{\infty}\dfrac{e^{n\pi}}{\pi^{ne}}$ **f** $\dfrac{3}{2}+\dfrac{6}{3}+\dfrac{9}{4}+\dfrac{12}{5}+\ldots$

> For part **f**, find the nth term.

g $\displaystyle\sum_{n=1}^{\infty}\dfrac{3}{2^{n-1}}$ **h** $\displaystyle\sum_{n=0}^{\infty}\left(\dfrac{1}{2^n}-\dfrac{1}{3^n}\right)$ **i** $\displaystyle\sum_{n=1}^{\infty}\dfrac{2^n}{2^{n+1}+1}$

2 Show that $\dfrac{1}{n^2-1}=\dfrac{1}{2(n-1)}-\dfrac{1}{2(n+1)}$, and hence determine if $\displaystyle\sum_{n=2}^{\infty}\dfrac{1}{n^2-1}$ converges.

3 Find the values of x for which $3+\dfrac{9}{x}+\dfrac{27}{x^2}+\ldots$ converges, and determine its sum in terms of x.

4.2 Introduction to convergence tests for series

From the previous section, we know that a *necessary* condition for a series to converge is that $\lim_{n\to\infty} u_n = 0$, but this is not a *sufficient* condition to ensure convergence. To understand what this means, we can use some real-life analogies. For example, breathing oxygen is a necessary condition for being alive, just as drinking liquids and having some basic nourishment are also necessary conditions for living, but they are by no means sufficient conditions. It is more difficult to define what conditions are *sufficient* for human life, as these will vary with environment, culture, gender, age (to name just a few), as well as our own individual definitions of what constitutes "living" and what makes life "worth living".

> In terms of the science of logic, the following statements are equivalent. Let A and B be statements, then
> - If A, then B
> - $A \Rightarrow B$
> - All A's are B's.
> - B is necessary for A.
> - A is sufficient for B.
> - $A \subseteq B$

We know that the *n*th test for divergence is necessary but not sufficient to ensure convergence. In this section we will be looking at several criteria that can be used to test for convergence of series. You already know one of these tests: the geometric series, and the conditions for which it will converge to its sum. We will begin, however, with a revision of some concepts from the core syllabus.

The integral as the limit of sums

In the following graph, consider the area under the curve $y = x^2$ from $x = 0$ to $x = 1$. The left-hand graph shows the actual area (shaded), and the right-hand graph shows an approximation of this area, using rectangles of base 0.125 and height x^2. You will notice that the error in our approximation is the total area of the white space between the curve and the rectangles. We can use the **method of exhaustion** to fill the space with more rectangles of smaller width.

> The method of exhaustion is a technique for finding the area of a shape by inscribing within it a sequence of polygons whose areas converge to the area of the containing shape. Antiphon of Athens (5th century BCE), Bryson of Heraclea (5th century BCE) and Archimedes (3rd century BCE) were the first to use and develop this method.

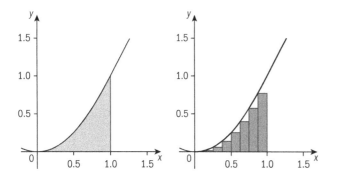

Using graphing software, it is easy to change *n*, the number of rectangles under the curve.

Using 15 similar rectangles, the approximation of the area under the curve is 0.3 sq. units.

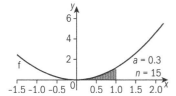 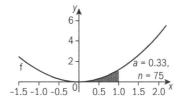 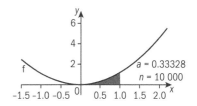

There is a better approximation when $n = 75$:

When $n = 10\,000$, the area is about 0.333 sq. units:

We can see from the above that if f is continuous in the interval $[a, b]$, to find the area under the curve of $f(x)$ from $x = a$ to $x = b$, we divide $[a, b]$ into n subintervals of equal length, $\frac{(b-a)}{n}$, and we call this Δx.

In each subinterval, we select the height of the rectangle such that a corner of the rectangle is on the curve, and call this $f(c)$.

You can see that as Δx approaches 0 (i.e. the number of rectangles n approaches infinity), the approximate area approaches the actual area.

Hence, the area under the curve of i such subintervals is approximated by $\sum_{i=1}^{n} f(c_i) \Delta x_i$, and is called a **Riemann sum** for the given function.

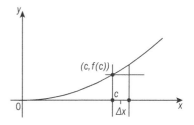

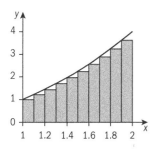

In the graph above we were considering rectangles below the curve, the so-called **lower bound sum**. We can also approximate the area by drawing rectangles above the curve, or the **upper bound sum**.

Again, we can consider the upper bound sum with 15, 75, and then 10 000 rectangles:

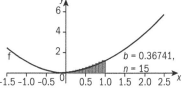
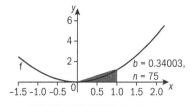
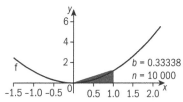

When $n = 15$, the area is approximately 0.367 sq. units

When $n = 75$, the area is approximately 0.340 sq. units.

When $n = 10{,}000$, the area is approximately 0.333 sq. units.

As we increase the number of rectangles, the approximate area will approach the actual area, and the combined area of i such subintervals above the curve is approximated by $\sum_{i=1}^{n} f(c_i)\Delta x_i$.

Hence, if the integral of f exists in the interval $[a, b]$, and f is non-negative in this interval, then the area, A, under the curve $y = f(x)$ from a to b is $A = \int_a^b f(x)dx$.

We can now define the definite integral as the limit of a Riemann sum.

Definition: Given that:

- f is a continuous function defined on the interval $[a, b]$;
- the interval $[a, b]$ is divided into n equal subintervals of width $\Delta x = \dfrac{b-a}{n}$;
- $x_i = a + i\Delta x$ is the right endpoint of the subinterval i;

then the **definite integral** of f from a to b is $\int_a^b f(x)dx$
$= \lim_{n \to \infty} \sum_{i=1}^{n} f(x_i)\Delta x.$

We have defined the definite integral as the limit of Riemann sums, as n goes to infinity.

> There are several internet applets that can help you understand Riemann sums, and they are far more effective than static screenshots. One such animation, where the number of rectangles can be increased in order to achieve better estimates of the area of under the curve, is:
> http://www.mathworld.wolfram.com/RiemannSum.html

Example 4

a For the function $f(x) = e^x$ and the partition of the interval $[0, 1]$ by $c_i = \dfrac{i-1}{n}, i = 1, 2, 3 \ldots n$ use a suitable Riemann sum to show that $\lim_{n \to \infty} \dfrac{1}{n}\left(1 + e^{\frac{1}{n}} + e^{\frac{2}{n}} + e^{\frac{3}{n}} + \ldots + e^{\frac{n-1}{n}}\right) = e - 1.$

b Use the Riemann sum definition of the definite integral to evaluate
$\int_0^4 (2x^2 + 3)\, dx$

a $\sum_{i=1}^{n} f(c_i) \Delta x_i = \sum_{i=1}^{n} e^{c_i} \frac{1}{n} = \sum_{i=1}^{n} e^{\frac{i-1}{n}} \frac{1}{n} = \frac{1}{n}\left(1 + e^{\frac{1}{n}} + e^{\frac{2}{n}} + e^{\frac{3}{n}} + \ldots + e^{\frac{n-1}{n}}\right)$	Definition of Riemann sum
Hence, $\lim_{n \to \infty} \frac{1}{n}\left(1 + e^{\frac{1}{n}} + e^{\frac{2}{n}} + e^{\frac{3}{n}} + \ldots + e^{\frac{n-1}{n}}\right) = \int_0^1 e^x \, dx = e - 1$	Definition of definite integral
b $\Delta x = \frac{4-0}{n} = \frac{4}{n}$	Divide [0, 4] into n equal subintervals.
$x_i = 0 + i\left(\frac{4}{n}\right) = \left(\frac{4i}{n}\right); \; f(x_i) = 2\frac{(4i)^2}{n^2} + 3 = \frac{32i^2}{n^2} + 3$ $f(x_i)\Delta x = \sum_{i=1}^{n}\left(\frac{32i^2}{n^2} + 3\right)\left(\frac{4}{n}\right) = \sum_{i=1}^{n}\left(\frac{128i^2}{n^3} + \frac{12}{n}\right)$	Define x_i, $f(x_i)$, and $f(x_i)\Delta x$
$\sum_{i=1}^{n}\left(\frac{128i^2}{n^3} + \frac{12}{n}\right) = \frac{128}{n^3}\sum_{i=1}^{n} i^2 + \frac{12}{n}\sum_{i=1}^{n} 1$ $= \frac{128}{n^3} \cdot \frac{n(n+1)(2n+1)}{6} + \frac{12}{n} \cdot n = \frac{128}{6} \cdot \frac{n}{n} \cdot \frac{n+1}{n} \cdot \frac{2n+1}{n} + 12$ $= \frac{64}{3}\left(1 + \frac{1}{n}\right)\left(2 + \frac{1}{n}\right) + 12$	Evaluate the sum
$\int_0^4 (2x^2 + 3)\, dx = \lim_{n \to \infty}\left[\frac{64}{3}\left(1 + \frac{1}{n}\right)\left(2 + \frac{1}{n}\right) + 12\right] = \frac{164}{3}$	Take the limit to infinity of the sum

We are now ready to formalize one of the most astonishing and important results in the development of calculus: the connection between the derivative and the definite integral. The following theorem justifies the procedures for evaluating definite integrals (such as areas under a curve) and is still regarded as one of the most significant developments of modern-day mathematics.

Fundamental theorem of calculus (FTC):

If f is continuous on $[a, b]$, then the function defined by $g(x) = \int_a^x f(t)\, dt$, $a \leq x \leq b$, is an anti-derivative of f, i.e. $g'(x) = f(x)$ for $a < x < b$.

The proof of the FTC uses the Weierstrass Theorem, which we saw in chapter 3, which states that if f is continuous and closed in $[a, b]$, then f has both a maximum and minimum on $[a, b]$.

> The Weierstrass Theorem is also referred to as the Extreme Value Theorem.

FTC Proof:

If $x, x + h \in (a, b)$ then

$$g(x+h) - g(x) = \int_a^{x+h} f(t)dt - \int_a^x f(t)dt$$

$$= \left[\int_a^x f(t)dt + \int_x^{x+h} f(t)dt\right] - \int_a^x f(t)dt$$

$$= \int_x^{x+h} f(t)dt$$

and hence, for $h \neq 0$

$$\frac{g(x+h) - g(x)}{h} = \frac{1}{h}\int_x^{x+h} f(t)\, dt. \quad (1)$$

Since f is continuous on $[x, x + h]$, by the extreme value theorem, there exist real numbers u and v in $[x, x + h]$ such that $f(u) = m$ and $f(v) = M$, where m and M are the absolute minimum and maximum respectively of f on $[x, x + h]$.

Hence, $m(b - a) \leq \int_a^b f(x)\, dx \leq M(b - a)$ and we have

$$mh \leq \int_x^{x+h} f(t)\, dt \leq Mh \Rightarrow f(u)\cdot h \leq \int_x^{x+h} f(t)\, dt \leq f(v)\cdot h.$$

Since $h > 0$, dividing the above by h gives us

$f(u) \leq \dfrac{1}{h}\displaystyle\int_x^{x+h} f(t)dt \leq f(v)$. Using the substitution (1) from above,

$$f(u) \leq \frac{g(x+h) - g(x)}{h} \leq f(v).$$

If we now let $h \to 0$, then $u \to x$ and $v \to x$, since both u and v lie between x and $x + h$. Hence,

$$\lim_{h \to 0} f(u) = \lim_{u \to x} f(u) = f(x); \lim_{h \to 0} f(v) = \lim_{v \to x} f(v) = f(x) \text{ since } f \text{ is}$$

continuous at x. Hence, by the squeeze theorem,

$$g'(x) = \lim_{h \to 0} \frac{g(x+h) - g(x)}{h} = f(x)$$

We will now state some corollaries of the FTC:

- $\displaystyle\int_x^a f(t)dt = -\int_a^x f(t)dt$

- $\displaystyle\int_a^b f(x)dx = F(b) - F(a)$

> This is sometimes also referred to as the fundamental theorem of calculus.

> The **fundamental theorem of calculus** was formalized and proven by **Augustin-Louis Cauchy (1789–1857)**. His proof elegantly joined the two branches of calculus; differential and integral calculus. Cauchy's last words before he died were indeed self-prophetic: *"Men pass away, but their deeds abide"*.

The finite in the infinite

Example 5

a If $F(x) = \int_x^3 \sqrt{1+t^{16}} \, dt$, find $F'(x)$.	
b Given that $5x^3 + 40 = \int_c^x f(t)dt$, find $f(x)$ and the value of c.	

a $F(x) = \int_x^3 \sqrt{1+t^{16}} \, dt = -\int_3^x \sqrt{1+t^{16}} \, dt$	Use one of the integral properties to get it into the FTC form.
$F'(x) = -\sqrt{1+x^{16}}$	Use the FTC.
b $\frac{d}{dx}[5x^3 + 40] = \frac{d}{dx}\left[\int_c^x f(t) \, dt\right]$	Take the derivative of both sides.
$15x^2 = f(x)$	Differentiate both sides, and use the definition of FTC
$5x^3 + 40 = \int_c^x 15t^2 \, dt$	Substitution
$\int_c^x 15t^2 \, dt = \left[5t^3\right]_c^x = 5x^3 - 5c^3$	Work out the RHS.
$5x^3 + 40 = 5x^3 - 5c^3$	Equate now LHS with above, and solve for c.
$40 = -5c^3$, hence $c = -2$.	

Exercise 4B

1. Prove the FTC corollaries.

2. For the function $f(x) = \sin(x)$ and the partition of the interval $[0, \pi]$ by intervals $c_i = \frac{i-1}{n}\pi$, use a suitable Riemann sum to

   show that $\lim_{n \to \infty} \left[\frac{1}{n}\left(\sin\frac{\pi}{n} + \sin\frac{2\pi}{n} + \sin\frac{3\pi}{n} + \ldots + \sin\frac{(n-1)\pi}{n}\right)\right] = \frac{2}{\pi}$

3. Use the Riemann sum definition of the definite integral to evaluate

   a $\int_0^5 2x \, dx$ **b** $\int_{-2}^0 (3x^2 + 2x) \, dx$

4. Use the FTC to find the derivative of

   a $\int_1^x \sqrt{3+4t} \, dt$ **b** $\int_x^4 2\tan t \, dt$ **c** $\frac{d}{dx}\int_x^0 t\sec t \, dt$

5. Let $F(x) = \int_4^x \sqrt{t^2 + 9t} \, dt$. Find

   a $F(4)$ **b** $F'(4)$ **c** $F''(4)$

6. If $G(x) = \int_0^x \sin(t) \, dt$, find $G(x)$ and $G'(x)$.

7. Find the interval on which $y = \int_0^x \frac{1}{1+t+t^2} \, dt$ is concave up.

The integral as the limit of a sum
4.3 Improper Integrals

The definition of the integral $\int_a^b f(x)\,dx$ requires that for the interval $[a, b]$, a and b must be real numbers. Furthermore, the fundamental theorem of calculus requires that f be continuous on the closed interval $[a, b]$. Integrals that do not meet these requirements are called **improper**.

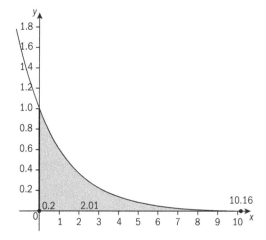

Definition: An improper integral $\int_a^b f(x)\,dx$ has infinity as one or both of its endpoints, or contains a finite number of infinite discontinuities in the interval $[a, b]$.

Consider the graph of the function $y = e^{\frac{-x}{2}}$ in the first quadrant.

It appears that the shaded region has infinite area, since $y = 0$ is an asymptote. In fact, it turns out that this area is finite! We will use integral notation to find the area under the curve, but since one of the endpoints is infinity, we have an improper integral.

To evaluate this integral, we proceed as follows, where $A(b)$ is the area under the curve from $a = 0$ to $b = \infty$.

$$A(b) = \int_0^b e^{\frac{-x}{2}}\,dx = \left[-2e^{\frac{-x}{2}}\right]_0^b = -2e^{\frac{-b}{2}} + 2$$

Now, find the limit of $A(b)$ as b approaches infinity, i.e.

$$\lim_{b\to\infty} A(b) = \lim_{b\to\infty}\left(-2e^{\frac{-b}{2}} + 2\right) = 2$$

Hence, the area under the curve from $a = 0$ to $b = \infty$ is

$$\int_0^\infty e^{\frac{-x}{2}}\,dx = \lim_{b\to\infty}\int_0^b e^{\frac{-x}{2}}\,dx = 2$$

Improper integral notation

If $f(x)$ is continuous on $[a, \infty[$, then $\int_a^\infty f(x)\,dx = \lim_{b\to\infty}\int_a^b f(x)\,dx$.

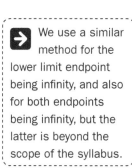

We use a similar method for the lower limit endpoint being infinity, and also for both endpoints being infinity, but the latter is beyond the scope of the syllabus.

Convergence of an improper integral

If the limit exists, the improper integral converges to the limiting value. If the limit fails to exist, the improper integral diverges.

The finite in the infinite

> The mathematician and physicist **Evangelista Torricelli (1608–47)** studied under Galileo, and is renowned for first writing about Torricelli's trumpet (also known as Gabriel's horn); an 'object' which has infinite surface area but finite volume! It is formed by the graph of the function $y = \dfrac{1}{x}$ rotated 2π radians about the x-axis.
>
>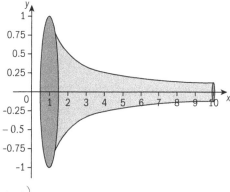
>
> Torricelli lived before calculus was discovered, so he came upon this paradox intuitively. When calculus was formalized, his result could be confirmed using improper integrals. If $V(b)$ is the volume of the 'trumpet' from 1 to infinity, then
>
> $V(b) = \pi \displaystyle\int_1^\infty \dfrac{1}{x^2} dx = \pi \lim_{b \to \infty} \int_1^b \dfrac{1}{x^2} dx = \pi \lim_{b \to \infty}\left[-\dfrac{1}{x}\right]_1^b = \pi \lim_{b \to \infty}\left(-\dfrac{1}{b} - (-1)\right) = \pi$
>
> Hence the integral converges to π.

Example 6

Evaluate, if possible, the following improper integrals by discussing convergence.

a $\displaystyle\int_1^\infty \dfrac{1}{x} dx$ **b** $\displaystyle\int_0^\infty \dfrac{1}{1+x^2} dx$

---

a $\displaystyle\int_1^\infty \dfrac{1}{x} dx = \lim_{b \to \infty} \int_1^b \dfrac{1}{x} dx$ *Convert the improper integral using appropriate notation.*

$= \lim_{b \to \infty}[\ln x]_1^b = \lim_{b \to \infty}(\ln b - 1) = \infty$ *Integrate the expression and find the limit, if it exists.*

The integral diverges.

b $\displaystyle\int_0^\infty \dfrac{1}{1+x^2} dx = \lim_{b \to \infty} \int_0^b \dfrac{1}{1+x^2} dx$ *Convert the improper integral using appropriate notation.*

$= \lim_{b \to \infty}[\arctan x]_0^b$ *Integrate the expression.*

$= \lim_{b \to \infty}(\arctan b - \arctan 0) = \dfrac{\pi}{2}$ *Find the limit if it exists.*

Thus, $\displaystyle\int_0^\infty \dfrac{1}{1+x^2} dx = \dfrac{\pi}{2}$, i.e. the integral converges to $\dfrac{\pi}{2}$.

Exercise 4C

Discuss the convergence of the following improper integrals, and evaluate if possible.

1 $\displaystyle\int_0^\infty e^{-x} dx$

2 $\displaystyle\int_0^\infty \sin x \, dx$

3 $\displaystyle\int_0^\infty \dfrac{e^x}{1+e^{2x}} dx$

4 $\displaystyle\int_1^\infty \dfrac{1}{\sqrt{x}} dx$

5 $\displaystyle\int_0^\infty \cos \pi x \, dx$

6 $\displaystyle\int_0^\infty x^2 e^{-x} dx$

Activity 1
Consider the definite integral $\int_0^1 \frac{1}{x^p} dx$.
1. By considering values of $p > 0$, explain why this integral is improper.
2. Show that the integral diverges if $p = 1$.
3. Show that the integral diverges if $p > 1$.
4. Show that the integral converges if $0 < p < 1$.

4.4 Integral test for convergence

There are similarities between infinite series and improper integrals of functions. We have seen that in the example of Torricelli's trumpet, $\int_0^\infty \frac{1}{x^2} dx = \lim_{b \to \infty} \int_1^b \frac{1}{x^2} dx = 1$.

We can compare the function $\frac{1}{x^2}$ to the series

$$\sum_{n=1}^\infty \frac{1}{n^2} = 1 + \frac{1}{4} + \frac{1}{9} + \ldots + \frac{1}{n^2} + \ldots$$

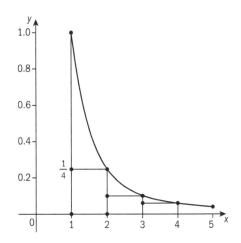

The rectangles above have areas $\frac{1}{2^2}, \frac{1}{3^2}, \frac{1}{4^2}, \ldots,$.

The sum of these areas is finite and less than the finite area under the curve. Hence we can say that the series convergences. This illustrates the following general principle:

Integral test: Let $f(x)$ be a continuous, positive, and decreasing function. Then the series $f(1) + f(2) + \ldots + f(n) + \ldots$ converges if the improper integral $\int_1^\infty f(x) dx$ converges, and diverges if the integral diverges.

> → The proof of the Integral test uses concepts beyond the scope of the course.

We can use the integral test to determine if the series

$$\frac{1}{2 \ln 2} + \frac{1}{3 \ln 3} + \frac{1}{4 \ln 4} + \ldots + \frac{1}{n \ln n} + \ldots \text{ converges or diverges.}$$

We would first examine the necessary condition for convergence, that is, $\lim_{n \to \infty} \frac{1}{n \ln n} = 0$, but we know that this alone is not sufficient to determine convergence.

Since $f(x) = x \ln x$ is increasing for $x \geq 2$, the function $f(x) = \frac{1}{x \ln x}$ is positive and decreasing.

We have the conditions necessary to use the integral test for convergence, therefore

$$\int_2^\infty \frac{1}{x \ln x} dx = \lim_{b \to \infty} \int_2^b \frac{1}{x \ln x} dx.$$

We can integrate using the method of substitution. Let $u = \ln x$, then $\frac{du}{dx} = \frac{1}{x}$ hence

$$\lim_{b \to \infty} \int_2^b \frac{1}{x \ln x} dx = \lim_{b \to \infty} \int_2^b \frac{1}{u} du = \lim_{b \to \infty} [\ln u]_2^b = \lim_{b \to \infty} (\ln(\ln b)) - \ln \ln 2 = \infty$$

Since the integral diverges, the series will also diverge.

Example 7

Determine if $\sum_{n=1}^{\infty} \frac{1}{n\sqrt{n}}$ converges or diverges.	
$\lim_{n \to \infty} \frac{1}{n\sqrt{n}} = 0$	The nth term test for divergence is inconclusive. The function is continuous, positive, and decreasing for all $x \geq 1$, hence we can apply the integral test.
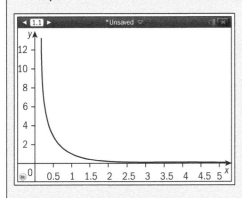	
$\int_1^\infty \frac{1}{x\sqrt{x}} dx = \lim_{b \to \infty} \int_1^b \frac{1}{x\sqrt{x}} dx$	Write the improper integral using the limit notation.
$\lim_{b \to \infty} \int_1^b \frac{1}{x\sqrt{x}} dx = \lim_{b \to \infty} \left[-2x^{\frac{-1}{2}}\right]_1^b$	Evaluate the limit.
$= \lim_{b \to \infty} \left(-\frac{2}{\sqrt{b}} + 2\right) = 2$	
The series converges.	Since the integral converges, so must the series.

A word of caution here is necessary. The series and the integral do not need to have the same value in the convergent case. In Example 7, the integral converges to 2. The series, however, can have quite a different sum. Use your GDC to calculate the partial sums of the series. You will notice that after the 11th partial sum the series is already greater than 2. Using the integral test does not necessarily tell you the sum of the series, but instead it will tell you whether or not the series converges.

Exercise 4D

Examine the necessary conditions to use the integral test, and determine whether the following series converge or diverge.

1. $\sum_{n=1}^{\infty} \dfrac{3}{n+1}$
2. $\sum_{n=1}^{\infty} \dfrac{n}{n^2+1}$
3. $\sum_{n=1}^{\infty} \dfrac{1}{n^2+1}$
4. $\sum_{n=1}^{\infty} n\sin\left(\dfrac{1}{n}\right)$
5. $\sum_{n=0}^{\infty} \dfrac{e^n}{1+e^{2n}}$
6. $\sum_{n=1}^{\infty} \dfrac{\cos n}{n^2}$
7. $\sum_{n=1}^{\infty} \dfrac{4n}{(2n^2+3)^2}$
8. $\sum_{n=1}^{\infty} \dfrac{n-\sqrt{n}}{n}$
9. $\sum_{n=1}^{\infty} \dfrac{2n+3}{\sqrt{n}}$
10. $\sum_{n=2}^{\infty} \dfrac{1}{n \ln n}$
11. $\sum_{n=1}^{\infty} \dfrac{\arctan n}{n^2+1}$
12. $\sum_{n=1}^{\infty} \dfrac{\ln n}{n^2}$
13. $\sum_{n=1}^{\infty} \dfrac{n^2}{e^n}$

4.5 The p-series test

The integral test can be used to determine the convergence of series of the form $\sum_{n=1}^{\infty} \dfrac{1}{n^p}$, where p is a real constant. (If we refer back to Example 5, we notice that the series is of this form, where $p = \dfrac{3}{2}$, and the series converged.) Such a series is called a *p*-series.

Activity 2

The *p*-series test

Using the integral test, prove that

1. $\sum_{n=1}^{\infty} \dfrac{1}{n^p}$ converges if $p > 1$
2. $\sum_{n=1}^{\infty} \dfrac{1}{n^p}$ diverges if $p < 1$
3. $\sum_{n=1}^{\infty} \dfrac{1}{n^p}$ diverges if $p = 1$

You will notice that the *p*-series is the harmonic series when $p = 1$.

Example 8

Determine the convergence or divergence of the following *p*-series. **a** $\sum_{n=1}^{\infty} \dfrac{1}{\sqrt[3]{n}}$ **b** $2 + \dfrac{2}{4} + \dfrac{2}{9} + \dfrac{2}{16} + \ldots$	
a $\sum_{n=1}^{\infty} \dfrac{1}{\sqrt[3]{n}} = \sum_{n=1}^{\infty} \dfrac{1}{n^{\frac{1}{3}}}$	Change the surd into a rational exponent
$p = \dfrac{1}{3}, \dfrac{1}{3} < 1$ hence the series diverges.	Identify p, and use the *p*-series test.
b The *n*th term of the series is $\dfrac{2}{n^2}$.	Identify the *n*th term of the series.
$S = \sum_{n=1}^{\infty} \dfrac{2}{n^2} = 2 \sum_{n=1}^{\infty} \dfrac{1}{n^2}$	Since $p = 2$, the series converges by the *p*-series.

Exercise 4E

Determine the convergence or divergence of the following series.

1 $\sum_{n=1}^{\infty} \dfrac{1}{n^\pi}$ **2** $\sum_{n=1}^{\infty} \dfrac{1}{n^{\frac{\pi}{4}}}$ **3** $1 + \dfrac{1}{2\sqrt{2}} + \dfrac{1}{3\sqrt{3}} + \dfrac{1}{4\sqrt{4}} + \ldots$

4 $1 + \dfrac{1}{\sqrt[5]{4}} + \dfrac{1}{\sqrt[5]{9}} + \dfrac{1}{\sqrt[5]{16}} + \ldots$ **5** $\sum_{n=1}^{\infty} \dfrac{2}{\sqrt{n}}$

We will now consider an effective way to show that a series of *non-negative* terms converges.

4.6 Comparison test for convergence

In Chapter 1 you learned the definition of boundedness for sequences. Another useful definition when working with sequences is the definition of **montonocity**.

Definition: A sequence $\{a_n\}$ is **monotonic** if its terms are non-decreasing,
i.e. $a_1 \leq a_2 \leq a_3 \leq \ldots \leq a_n \ldots$ or if it terms are non-increasing,
i.e. $a_1 \geq a_2 \geq a_3 \geq \ldots \geq a_n \ldots$

Both of these definitions are necessary to state the following theorem.

Theorem 3: If a sequence is both bounded and monotonic then it converges.

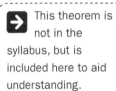

→ This theorem is not in the syllabus, but is included here to aid understanding.

The diagram below shows a monotonic increasing sequence that is bounded above, hence it converges.

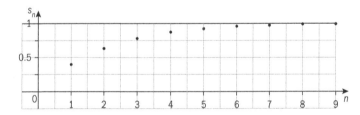

(As an exercise, you may want to try to prove this theorem using the definition of a convergent sequence from Chapter 1.)

Using this theorem, we can state the following theorem that enables us to test convergence or divergence of a given series by comparing it to one whose convergence or divergence is known.

Theorem 4 (Comparison Test):

Let $\sum a_n$ and $\sum b_n$ be series with non-negative terms, and $0 \leq a_n \leq b_n$ for all n. Then

1 If $\sum b_n$ converges, then $\sum a_n$ converges.

2 If $\sum a_n$ diverges then $\sum b_n$ also diverges.

Proof:

Let $L = \sum b_n$; $S_n = a_1 + a_2 + a_3 + \ldots$.

Since $0 \leq a_n \leq b_n$, the sequence $S_1, S_2, S_3, \ldots$ is monotonic, and bounded above by L, hence it converges. Furthermore, since $\lim_{n \to \infty} S_n = \sum a_n$ it follows that $\sum a_n$ converges. The second property follows since it is the contrapositive of the first property.

It is important to remember that this test, as well as the tests we've done until now, holds from some point on, not necessarily starting at $n = 1$. For example, if the series $\sum a_n$ converges, and if $b_n \leq a_n$ only for $n \geq 500$ then the series $\sum b_n$ also converges. The initial finite sum of the first 499 terms of $\{b_n\}$ can be any value, as long as ultimate behavior of the converges or diverges. You will see examples of this later on in the chapter.

A very useful family of series to use in applying the comparison test is the *p*-series. We can now prove the following theorem, which we actually did in Activity 2.

Theorem 5: The series $\sum \dfrac{1}{n^p}$ diverges if $p \leq 1$ and converges if $p > 1$.

For $0 < p \leq 1$ $\dfrac{1}{n^p} \geq \dfrac{1}{n}$, and so the series diverges by comparison with the harmonic series $\sum \dfrac{1}{n}$, which we know to diverge.

For $p > 1$, the partial sums of the series are always less than the area under the curve of the function $y = \dfrac{1}{x^p}$, as seen below in the graph for $p = 2$.

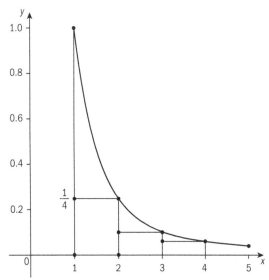

We say that the sequence of partial sums of the series is *bounded* by the area under the curve, or is always less than or equal to the area under the curve, and hence the series converges.

Example 9

Determine if the following series converge or diverge.

a $\sum \dfrac{\sin^2 n}{3^n}$ **b** $\sum \dfrac{n}{3n+1}$

---

a $0 < \sin^2 n \leq 1$ hence $\dfrac{\sin^2 n}{3^n} \leq \dfrac{1}{3^n}$

$\sum \dfrac{1}{3^n}$ converges.

Hence, the given series converges.

We look for a familiar series to compare the given series with.

It is a geometric series with $r = \dfrac{1}{3}$.

b $\dfrac{1}{n} < \dfrac{n}{3n+1}$ for $n > 4$

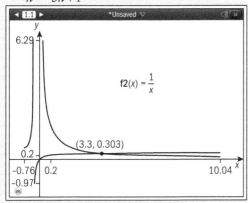

Hence, by the comparison test, the series diverges.

We look for a familiar series to compare the given series with. In this case it is the harmonic series, since it has the same exponent on n as the given series.

Do either a term-by-term comparison, or graph the functions represented by the series. Finding the point of intersection, you can clearly see that the given series is larger than the harmonic for all $n > 4$.

At times the comparison test may be inconclusive when using a familiar series. For example, determine whether or not the series $\sum \dfrac{1}{2+\sqrt{n}}$ converges.

We know that $\dfrac{1}{\sqrt{n}}$ diverges by the *p*-series test. Furthermore, $\dfrac{1}{2+\sqrt{n}} \leq \dfrac{1}{\sqrt{n}}$, for all $n \geq 1$. This, however, does not satisfy the requirements for using the comparison test, since our given series is *smaller* than a divergent series. Thus here, the comparison test has been inconclusive, so we cannot say whether the series converges or diverges. Instead, let's compare our given series to the harmonic series, then $\dfrac{1}{n} \leq \dfrac{1}{2+\sqrt{n}}$ for all $n \geq 4$. (Using the functions the series represent, and graphing them on the GDC, this result is evident.) Hence, we can conclude that the given series is divergent.

Exercise 4F

Determine if the series converge or diverge.

1 $\sum \dfrac{1}{n^3+1}$ **2** $\sum \dfrac{1}{\sqrt{n}-1}$ **3** $\sum \dfrac{1}{(n+1)!}$ **4** $\sum \dfrac{1}{\ln n}$ **5** $\sum \dfrac{1}{\sqrt{n^3+1}}$

6 $\sum \dfrac{\ln n}{n}$ **7** $\sum \left(\dfrac{\cos n}{n}\right)^2$ **8** $\sum \dfrac{n-1}{n^2\sqrt{n}}$ **9** $\sum \dfrac{\arctan n}{\sqrt{n^3+1}}$ **10** $\sum \dfrac{1}{n^{\ln n}}$

With some series it might not be possible to establish a term-by-term comparison needed to employ the comparison test. In such cases we need another test.

4.7 Limit comparison test for convergence

Theorem 6: Let $a_n > 0$ and $b_n > 0$ for all $n \geq N$. If $\lim\limits_{n \to \infty} \dfrac{a_n}{b_n} = c$, c a positive real number, then both $\sum a_n$ and $\sum b_n$ converge, or both diverge.

The limit comparison test allows us to prove that the general harmonic series $\sum\limits_{n=1}^{\infty} \dfrac{1}{an+b}$ diverges. We can compare it to the harmonic series and form the ratio

$$\lim_{n \to \infty} \dfrac{\dfrac{1}{an+b}}{\dfrac{1}{n}} = \lim_{n \to \infty} \dfrac{n}{an+b} = \lim_{n \to \infty} \dfrac{1}{a + \dfrac{b}{n}} = \dfrac{1}{a}.$$

Since $\dfrac{1}{a} > 0$, and the harmonic series diverges, then the general harmonic series also diverges.

This test works well with comparing series with the p-series.

When selecting a p-series, choose one whose nth term has the same order as the nth term of the given series. For example, given the series $\sum \dfrac{1}{\sqrt{5n-1}}$, select the series $\sum \dfrac{1}{\sqrt{n}}$ as comparison. Given $\sum \dfrac{2n^2 - 3}{3n^5 + n^3}$, select the series $\sum \dfrac{n^2}{n^5}$, or $\sum \dfrac{1}{n^3}$, as comparison.

Example 10

Determine if the following series converge or diverge by using the limit comparison test.

a $\sum \dfrac{n}{n^2 + 1}$ **b** $\sum \dfrac{2\sqrt{n} - 3}{3n^2 + 4\sqrt{n}}$

a Compare to $\sum \dfrac{n}{n^2} = \sum \dfrac{1}{n}$.

Select a p-series with the same order as that of the given series.

$$\lim_{n \to \infty} \dfrac{\dfrac{n}{n^2+1}}{\dfrac{n}{n^2}} = \lim_{n \to \infty} \left(\dfrac{n}{n^2+1} \cdot \dfrac{n^2}{n} \right) = \lim_{n \to \infty} \dfrac{n^2}{n^2+1} = 1$$

Find the limit of the ratio of the given series with the p-series.

Since there exists a positive limit, and the harmonic series diverges, then the given series diverges.

Note that this series was found to diverge using the integral test in Question 2 of Exercise 3C.

b Compare to $\sum \dfrac{\sqrt{n}}{n^2} = \sum \dfrac{1}{n^{\frac{3}{2}}}$.

$$\lim_{n \to \infty} \dfrac{\dfrac{2\sqrt{n}-3}{3n^2 + 4\sqrt{n}}}{\dfrac{1}{n^{\frac{3}{2}}}} = \lim_{n \to \infty} \dfrac{2\sqrt{n} - 3}{3n^2 + 4\sqrt{n}} \cdot n^{\frac{3}{2}} = \dfrac{2}{3}$$

Select a p-series with the same order as that of the given series.

Find the limit of the ratio of the given series with the p-series.

Since there exists a positive limit, and the p-series converges, then the given series converges.

Exercise 4G

Using the limit comparison test, determine the convergence or divergence of the following series.

1. $\sum_{n=0}^{\infty} \dfrac{1}{\sqrt{n^2+1}}$
2. $\sum_{n=1}^{\infty} \dfrac{2n^2-1}{4n^5+n-1}$
3. $\sum_{n=1}^{\infty} \dfrac{1}{n+\sqrt{n}}$
4. $\sum_{n=1}^{\infty} \dfrac{1}{2^n-1}$
5. $\sum_{n=1}^{\infty} \dfrac{3n^2+2n}{\sqrt{4+n^5}}$
6. $\sum_{n=1}^{\infty} \left(\dfrac{1}{2n-1}-\dfrac{1}{2n}\right)$
7. $\sum_{n=1}^{\infty} \dfrac{1}{2\sqrt{n}+\sqrt{n+2}}$
8. $\sum_{n=1}^{\infty} \dfrac{n2^n+1}{3n^2+2n}$

4.8 Ratio test for convergence

We have explored two tests for convergence of series with non-negative terms. Although useful, the drawback of these two tests is that we must find a familiar series that we know either diverges or converges in order to perform these tests.

We shall now consider the **ratio test**, which determines convergence without reference to any other series. Although we do not need any other series for comparison, the ratio test has a drawback – namely that it is inconclusive under certain circumstances. This test determines convergence by considering the ratio of successive terms of the series.

Ratio test: Let $a_n > 0$ for $n \geq 1$, and $\lim_{n\to\infty} \dfrac{a_{n+1}}{a_n} = L$. Then $\sum_{n=1}^{\infty} a_n$ converges if $L < 1$, and diverges if $L > 1$. If $L = 1$ the test is inconclusive.

> The ratio test is also referred to as the **d'Alembert ratio test** because it was first published by the French mathematician, physicist, and music theorist **Jean le Rond d'Alembert (1717–83)**.

Example 11

Determine the convergence or divergence of the following series.

a $\sum_{n=1}^{\infty} \dfrac{n}{2^n}$
b $\sum_{n=1}^{\infty} \dfrac{2^n}{n!}$

a $\lim_{n\to\infty} \dfrac{a_{n+1}}{a_n} = \dfrac{\frac{n+1}{2^{n+1}}}{\frac{n}{2^n}}$ — *Use the ratio test.*

$= \lim_{n\to\infty} \left(\dfrac{n+1}{n} \cdot \dfrac{2^n}{2^{n+1}}\right) = \dfrac{1}{2}\lim_{n\to\infty}\left(1+\dfrac{1}{n}\right) = \dfrac{1}{2}$ — *Find the limit of the ratio.*

Since $L < 1$ the series converges.

b $\lim_{n\to\infty} \dfrac{\frac{2^{n+1}}{(n+1)!}}{\frac{2^n}{n!}}$ — *Use the ratio test.*

$= \lim_{n\to\infty} \left(\dfrac{n!}{(n+1)!} \cdot \dfrac{2^{n+1}}{2^n}\right) = \lim_{n\to\infty} \dfrac{2}{n+1} = 0$ — *Find the limit of the ratio.*

Since $L < 1$ the series converges.

Using the ratio test, we will now consider two cases where $L = 1$.
In one case the series diverges, and in the other case it converges.

Let us apply the ratio test to the harmonic series, i.e. we need to evaluate $\lim_{n\to\infty} \dfrac{\frac{1}{n+1}}{\frac{1}{n}}$.

This equals $\lim_{n\to\infty} \dfrac{n}{n+1} = \lim_{n\to\infty} \dfrac{1}{1+\frac{1}{n}} = 1$.

However, we know that the harmonic series diverges.

Let us now apply the ratio test to the series $\sum_{n=1}^{\infty} \dfrac{1}{n^2}$.

$$\lim_{n\to\infty} \dfrac{\frac{1}{(n+1)^2}}{\frac{1}{n^2}} = \lim_{n\to\infty} \dfrac{n^2}{(n+1)^2} = \lim_{n\to\infty} \left(\dfrac{n}{n+1}\right)^2 = \lim_{n\to\infty} \left(\dfrac{1}{1+\frac{1}{n}}\right)^2 = 1.$$

However, we know that this series converges by the p-series test.

Hence, when $L = 1$, the ratio test is inconclusive, as we have seen examples now where the series can be divergent or convergent.

Exercise 4H

Using the Ratio test, determine whether the following series converge or diverge.

1. $\sum_{n=1}^{\infty} \dfrac{2^n}{n}$
2. $\sum_{n=1}^{\infty} \dfrac{1}{n!}$
3. $\sum_{n=1}^{\infty} \dfrac{3^n}{n!}$
4. $\sum_{n=1}^{\infty} \dfrac{3}{n2^n}$
5. $\sum_{n=1}^{\infty} \dfrac{\pi^n}{3n+2}$
6. $\sum_{n=1}^{\infty} \dfrac{5^{n+2}}{(n+1)!}$

4.9 Absolute convergence of series

All of the tests for convergence considered thus far assume that all the terms of the series are non-negative. This condition does not limit the usefulness of the tests. We can apply our test to $\sum |a_n|$, which of course has no negative terms.

Theorem 7: If the series $\sum |a_n|$ converges, then $\sum a_n$ converges, and we say that $\sum a_n$ converges absolutely.

We can prove this theorem using the comparison test. For each
n, $-|a_n| \leq a_n \leq |a_n|$, hence $0 \leq a_n + |a_n| \leq 2|a_n|$. If $\sum |a_n|$ converges then $2 \sum |a_n|$ will also converge, and by the comparison test, the
non-negative series $\sum (a_n + |a_n|)$ converges. Using the equality $a_n = (a_n + |a_n|) - |a_n|$, we can express $\sum |a_n|$ as the difference of two convergent series, i.e.
$\sum a_n = \sum (a_n + |a_n| - |a_n|) = \sum (a_n + |a_n|) - \sum |a_n|$. There $\sum a_n$ converges.

Example 12

Determine if the following series converge absolutely, and hence whether they converge.

a $1 + \dfrac{1}{2^2} - \dfrac{1}{3^2} + \dfrac{1}{4^2} - \dfrac{1}{5^2} + \ldots$ **b** $\displaystyle\sum_{n=1}^{\infty} \dfrac{(-1)^{n-1}}{\sqrt{n}}$

Both of the series above are alternating series, i.e. the terms of the series are alternately positive and negative. The series in part **c** below also has positive and negative terms, but it is not an alternating series.

c $\displaystyle\sum_{n=1}^{\infty} \dfrac{\cos n}{n^2}$ **d** $\displaystyle\sum_{n=1}^{\infty} (-1)^n \dfrac{n^3}{3^n}$

---

a This series is not non-negative, so we test for absolute convergence.	*Determine the nature of the series.*
The series of absolute values is less than or equal to $\sum \dfrac{1}{n^2}$ which converges by the *p*-series test. Thus by the comparison test, $\displaystyle\sum_{n=1}^{\infty} \dfrac{|\cos n|}{n^2}$ converges.	*The series is absolutely convergent.*
Since the series converges absolutely, the series converges.	
b $\displaystyle\sum_{n=1}^{\infty} \dfrac{(-1)^{n-1}}{\sqrt{n}} = 1 - \dfrac{1}{\sqrt{2}} + \dfrac{1}{\sqrt{3}} - \dfrac{1}{\sqrt{4}} + \ldots$	*Write out the terms of the series.*
The series of absolute values is $\sum \dfrac{1}{\sqrt{n}}$, which diverges by the *p*-series test. Hence, we cannot say if the series converges.	*The series is not absolutely convergent, i.e. it diverges absolutely.*
c $\displaystyle\sum_{n=1}^{\infty} \dfrac{\cos n}{n^2} = \dfrac{\cos 1}{1^2} + \dfrac{\cos 2}{2^2} + \dfrac{\cos 3}{3^2} + \ldots$ and has terms of different signs, since $-1 \leq \cos n \leq 1$.	*Write out some terms of the series.*
Using the comparison test, $\left\|\dfrac{\cos n}{n^2}\right\| \leq \dfrac{1}{n^2}$	*Choose an appropriate convergence test.*
$\displaystyle\sum_{n=1}^{\infty} \dfrac{1}{n^2}$ converges by the *p*-series test.	*Apply the convergence test.*
The series converges absolutely, hence the series converges.	*Apply the absolute convergence test.*
d Using the ratio test: $\displaystyle\lim_{n\to\infty} \dfrac{\dfrac{(n+1)^3}{3^{n+1}}}{\dfrac{n^3}{3^n}}$	*Choose an appropriate convergence test for the absolute value of the series. (No need to include the absolute value sign as this expression is always positive.)*
$= \displaystyle\lim_{n\to\infty}\left(\dfrac{(n+1)^3}{3^{n+1}} \cdot \dfrac{3^n}{n^3}\right) = \lim_{n\to\infty}\left[\dfrac{1}{3}\left(\dfrac{n+1}{n}\right)^3\right] = \dfrac{1}{3}$	*Evaluate the limit.*
Since $L < 1$, the series converges absolutely. Hence the series converges.	*Apply the ratio test.*

Exercise 4I

Determine if the following series converge absolutely, and hence whether they converge.

1 $\sum_{n=1}^{\infty} (-1)^n \dfrac{n}{2^n}$
2 $\sum_{n=1}^{\infty} \dfrac{\sin n}{n^2}$
3 $\sum_{n=1}^{\infty} \dfrac{(-2)^n}{n!}$

4 $\sum_{n=1}^{\infty} (-1)^n \dfrac{\arctan n}{n^3}$
5 $\sum_{n=0}^{\infty} \dfrac{(-1)^n}{2^n n!}$
6 $\sum_{n=0}^{\infty} \dfrac{(-1)^n n!}{e^n}$

4.10 Conditional convergence of series

We have seen in part **b** of Example 12 that the series is not absolutely convergent. This does not, however, imply that the series is divergent. From Theorem 7, we know that if the series converges absolutely, then it converges. If it does not converge absolutely, we need more information, or a different test, to determine convergence. In the case of an alternating series, a series whose terms are alternately positive and negative, we do indeed have more help to determine convergence.

Let us consider the harmonic alternating series, i.e.

$$\sum_{n+1}^{\infty} (-1)^{n+1} \dfrac{1}{n} = 1 - \dfrac{1}{2} + \dfrac{1}{3} - \dfrac{1}{4} + \ldots$$

We can clearly see that its terms are decreasing in absolute value, and that $\lim_{n \to \infty} a_n = 0$. In this case, we have Leibniz's theorem to help us.

Alternative series test (Leibniz's theorem)

The series $\sum_{n=1}^{\infty} (-1)^{n+1} u_n = u_1 - u_2 + u_3 - u_4 + \ldots$, where $u_n > 0$ for all n in $\mathbb{Z}^+$, converges if the following hold:

1 $u_n \geq u_{n+1}$ for all $n \geq N$, where N is a positive integer;
2 $\lim_{n \to \infty} u_n = 0$.

The proof of Leibniz's theorem is beyond the level of this course.

Using the Alternative series test, we can easily prove that the alternating harmonic series converges. The absolute value of its terms are decreasing, and their limit value as n approaches infinity is 0. We say that the alternating harmonic series *converges conditionally*.

Definition: The infinite series $\sum u_n$ is **conditionally convergent** if $\sum u_n$ converges but $\sum |u_n|$ diverges.

We can now come back to part **b** of Example 10, and determine if $\sum_{n=1}^{\infty} (-1)^{n-1} \dfrac{1}{\sqrt{n}}$ converges conditionally. We have already shown in the example that it diverges absolutely. It is also easy to see that for all $n \geq 1 \Rightarrow u_n > u_{n+1}$. Hence, the two conditions of Leibniz's theorem are satisfied, and the series converges conditionally.

Example 13

Determine if the series $\sum_{n=1}^{\infty} \frac{(-1)^n}{\ln(n+1)}$ converges or diverges. If the series converges, determine if it is absolutely or conditionally convergent.	
The absolute value of the terms is decreasing, and $\lim_{n \to \infty} u_n = 0$.	Test the conditions of Leibniz's theorem.
Hence, the series converges.	It meets the conditions of Leibniz's theorem.
For the series of absolute values, $\frac{1}{n} < \frac{1}{\ln(n+1)}$. Hence by the comparison test, since the harmonic series diverges, the series diverges.	Use an appropriate test for the convergence of the series of absolute values.
The refore, the series converges conditionally.	It satisfies the definition of conditional convergence.

The diagram on the right illustrates the convergence of partial sums to their limit S.

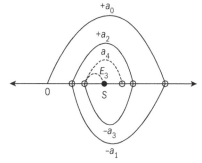

The illustration shows the way that the alternating series converges when it satisfies the conditions of Leibniz's theorem. The partial sums (circles on the number line) oscillate back and forth on the number line until they gradually close in on the limit S as the terms tend to 0. Stopping at any nth partial sum we know that the next term, $\pm u_{n+1}$, will again surpass the limit in either the positive or negative direction (E_3 on the graph). This gives us a convenient method for estimating what is called the truncation error of an alternating series.

Theorem 9: If the alternating series $\sum_{n=1}^{\infty} (-1)^{n+1} u_n$ satisfies the conditions to apply Leibniz's theorem, then the truncation error for the nth partial sum is less than u_{n+1}, and has the same sign as the unused term. The sum of the series S_∞ is therefore $S_n \pm R_n$ 'the truncation error', i.e. $|S_\infty - S_n| = |R_n| \leq u_{n+1}$, where R_n is the truncation error, or remainder.

We now know that the alternating harmonic series converges conditionally, and hence we can estimate the truncation error – for example, after the 100th term. We know that the error is less than u_{101} which is $\frac{1}{101}$. Using the GDC and obtaining exact answers, we calculate the sum of the first 100 terms, and the sum of the first 101 terms, and then calculate their difference. The result confirms our answer above.

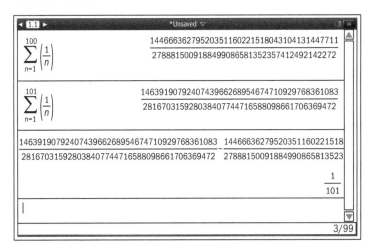

We can use this truncation error to estimate the sum of the series. In other words, $S_\infty = S_{100} \pm \dfrac{1}{101}$.

Example 14

Approximate the sum of the series $\sum_{n=1}^{\infty}(-1)^{n-1}\dfrac{1}{n!}$ using its first six terms.

$\dfrac{1}{n!} > \dfrac{1}{(n+1)!}$ for all n, and $\lim\limits_{n\to\infty}\dfrac{1}{n!} = 0$, hence the series converges.	Test conditions for Leibniz's theorem.
The truncation error is $u_7 = \dfrac{1}{7!} = \dfrac{1}{5040}$.	Since the conditions for Leibniz's theorem are met, the truncation error can be estimated.
Hence $S = S_6 + \dfrac{1}{5040} \approx 0.632$	The truncation error will have the same sign as the truncated term.

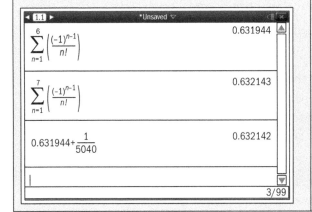

Example 15

Approximate $\sum_{n=1}^{\infty} \frac{(-1)^{n+1}}{2n^3-1}$ with an error of less than 0.001.

$u_{n+1} < u_n$ for all $n \geq 1$, $\lim_{n\to\infty} \frac{1}{2n^3-1} = 0$ The series converges.	Check conditions for convergence of alternating series test.
$u_{n+1} \leq 0.001 \Rightarrow \frac{1}{2(n+1)^3 - 1} \leq 0.001;$ $n = 7$ $S_7 \approx 0.947.$	Apply the truncation error theorem. See GDC screenshot below. Find S_7. See GDC screenshot below.

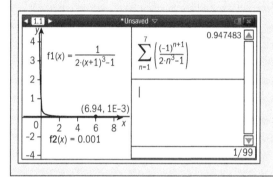

Exercise 4J

1 Determine if the following series converge or diverge. If the series converges, determine if it is absolutely or conditionally convergent.

  a $\sum_{n=2}^{\infty} (-1)^n \frac{1}{\ln n}$ **b** $\sum_{n=1}^{\infty} \frac{(-1)^n n}{2^n}$ **c** $\sum_{n=1}^{\infty} \frac{(-1)^n}{\sqrt{n}}$ **d** $\sum_{n=0}^{\infty} \frac{(-1)^n}{n!}$

  e $\sum_{n=0}^{\infty} (-1)^n e^{-n}$ **f** $\sum_{n=1}^{\infty} \frac{\cos n\pi}{n}$ **g** $\sum_{n=2}^{\infty} \frac{(-1)^{n-1}}{n \ln n}$ **h** $\sum_{n=1}^{\infty} \frac{(-1)^{n-1} \arctan\left(\frac{1}{n}\right)}{n^2}$

2 Approximate the sum of the series $\sum_{n=1}^{\infty} \frac{(-1)^{n-1}}{n^3}$ using its first ten terms, to 3 d.p.

3 Approximate the sum of the series $\sum_{n=1}^{\infty} \frac{(-1)^{n+1}}{n^4}$ with an error of less than 0.001.

4 Approximate the sum of the series $\sum_{n=2}^{\infty} \frac{(-1)^{n+1}}{n^3 3^n}$ to 6 d.p., using its first three terms, and obtain an estimate of its truncation error.

5 Find the least number of terms necessary to approximate the sum of $\sum_{n=2}^{\infty} \frac{(-1)^{n+1}}{n^3 \ln n}$ with an error less than 10^{-5}.

Summary of tests for convergence

Test	Series	Convergence or Divergence	Comments
nth term	$\sum a_n$	Diverges if $\lim_{n\to\infty} a_n \neq 0$	If $\lim_{n\to\infty} a_n = 0$, test is inconclusive
Geometric series	$\sum_{n=1}^{\infty} ar^{n-1}$	Converges to sum, if $S_\infty = \dfrac{a}{1-r}$, $|r|<1$. Diverges otherwise.	Useful for comparison tests.
p-series	$\sum_{n=1}^{\infty} \dfrac{1}{n^p}$	Converges when $p > 1$, otherwise diverges.	Useful for comparison tests.
Integral	$\sum_{n=1}^{\infty} a_n;\ a_n = f(n)$	Converges if $\int_1^\infty f(x)dx$ converges; diverges if $\int_1^\infty f(x)dx$ diverges.	$f(x)$ must be continuous, positive, and decreasing.
Comparison	$\sum a_n, \sum b_n;\ a_n \geq 0, b_n \geq 0$	If $\sum b_n$ converges and $a_n \leq b_n$ for all n then $\sum a_n$ converges. If $\sum b_n$ diverges and $a_n \geq b_n$ for all n, then $\sum a_n$ diverges.	The comparison series is often geometric or a p-series.
Limit comparison	$\sum a_n, \sum b_n;\ a_n \geq 0, b_n \geq 0$	If $\lim_{n\to\infty}\left(\dfrac{a_n}{b_n}\right) = c,\ c \in R^+$, then both converge or both diverge.	To find b_n consider only terms of a_n that have the greatest effect on the magnitude.
Ratio	$\sum a_n$	If $\lim_{n\to\infty}\left|\dfrac{a_{n+1}}{a_n}\right| = L$, the series converges (absolutely) if $L < 1$, and diverges otherwise.	Test is inconclusive if $L = 1$.
Alternating	$\sum (-1)^n a_n,\ a_n > 0,$	Converges if $a_k \geq a_{k+1}$ for all k, and $\lim_{n\to\infty} a_n = 0$.	Only applicable to alternating series.
$\sum |a_n|$	$\sum a_n$	$\sum |a_n| \Rightarrow \sum a_n$ converges.	If $\sum a_n$ converges, but $\sum |a_n|$ diverges, then $\sum a_n$ converges conditionally.

Exercise 4K

Using any appropriate test, determine if the following series converge or diverge. For alternating series, determine if they converge absolutely or conditionally.

1. $\sum_{n=1}^{\infty} \dfrac{6^n}{n!}$

2. $\sum_{n=1}^{\infty} \dfrac{n2^n}{3^n}$

3. $\sum_{n=0}^{\infty} e^{-n}$

4. $\sum_{n=1}^{\infty} \dfrac{1}{n^2} \tan\left(\dfrac{1}{n}\right)$

5. $\sum_{n=1}^{\infty} \dfrac{2n+1}{2n^2+3n-1}$

6. $\sum_{n=1}^{\infty} \dfrac{(-1)^{n-1}}{\sqrt{3n-1}}$

7. $\sum_{n=2}^{\infty} \dfrac{2}{\ln n^2}$

8. $\sum_{n=1}^{\infty} \dfrac{n(n+1)}{\sqrt{n^3+5n^2}}$

9. $\sum_{n=2}^{\infty} \dfrac{(-1)^{n+1}}{n\sqrt{\ln n}}$

10. $\sum_{n=1}^{\infty} \dfrac{4^n n^2}{n!}$

11. $\sum_{n=0}^{\infty} \dfrac{n!}{2 \cdot 5 \cdot 8 \cdot \ldots \cdot (3n+2)}$

12. $\sum_{n=1}^{\infty} \dfrac{(-2)^{2n}}{n^n}$

13. $\sum_{n=1}^{\infty} \dfrac{\sin\left(\frac{1}{n}\right)}{\sqrt{n}}$

14. $\sum_{n=1}^{\infty} \dfrac{\arctan n}{n^{\frac{3}{2}}}$

15. Explain why the indicated test for convergence cannot be used with the given series.

   a Integral test; $\sum_{n=1}^{\infty} e^{-n} \sin n$

   b Comparison test; $\sum_{n=1}^{\infty} \dfrac{(-1)^n}{n^2}$

   c Ratio test; $\sum_{n=1}^{\infty} \sin n$

   d Alternating series: i $\sum_{n=1}^{\infty} (-1)^{n-1} n$ ii $\sum_{n=1}^{\infty} (-1)^{n-1}\left(2 - \dfrac{1}{n}\right)$

16. a By finding the first six partial sums of the series $\sum_{n=1}^{\infty} \dfrac{n}{(n+1)!}$, conjecture a formula for S_n and use mathematical induction to prove your conjecture.

   b Show that the series $\sum_{n=1}^{\infty} \dfrac{n}{(n+1)!}$ is convergent and find its sum.

Review exercise

EXAM-STYLE QUESTIONS

1. Determine if the following series converge or diverge.

   a $\sum_{n=0}^{\infty} \left(\dfrac{n}{n+4}\right)^n$
   b $\sum_{n=1}^{\infty} \sin \dfrac{1}{n}$
   c $\sum_{n=2}^{\infty} \dfrac{1}{\ln n}$
   d $\sum_{n=1}^{\infty} \dfrac{n}{e^n}$
   e $\sum_{n=1}^{\infty} (n+10)\left(\dfrac{\cos n\pi}{n^{1.4}}\right)$

2. Show that the series $\sum_{k=1}^{\infty} \dfrac{(k-1)}{k!}$ is convergent, and find the sum of the series.

3. Prove that the series $\sum_{n=1}^{\infty} \dfrac{(-1)^n}{(n+1)^7}$ converges, and approximate its sum to 6 d.p.

4. Show that the series $\sum_{n=1}^{\infty} \dfrac{1+n}{1+n^2}$ diverges.

EXAM-STYLE QUESTIONS

5 Let $\sum_{n=1}^{\infty} a_n$ and $\sum_{n=1}^{\infty} b_n$ be two convergent series of positive terms.

  a Show that $a_n \cdot b_n < b_n$ for large values of n.

  b Show that $\sum_{n=1}^{\infty} a_n b_n$ converges, and hence show that $\sum_{n=1}^{\infty} a_n^2$ converges.

6 Determine whether or not the following series converge.

  a $\dfrac{1}{\sqrt[3]{2}} + \dfrac{4}{\sqrt[3]{4}} + \dfrac{7}{2} + \dfrac{10}{2\sqrt[3]{2}} + \ldots$ **b** $\dfrac{1}{\sqrt{2}} + \dfrac{3}{2} + \dfrac{5}{2\sqrt{2}} + \dfrac{7}{4} + \dfrac{9}{4\sqrt{2}} + \ldots$

7 Using the limit comparison test with $\sum \dfrac{1}{n^2}$, determine if $\sum 1 - \cos\dfrac{1}{n}$ converges or diverges.

8 a Draw a suitable sketch to show that for

  $p > 1$, $\sum_{n=2}^{\infty} \dfrac{1}{n^p} < \int_1^{\infty} \dfrac{1}{x^p} dx < \sum_{n=1}^{\infty} \dfrac{1}{n^p}$.

  b Hence, show that for $p > 1$, $\dfrac{1}{p-1} \leq \sum_{n=1}^{\infty} \dfrac{1}{n^p} \leq \dfrac{p}{p-1}$.

9 a The diagram below shows a sketch of the graph $y = x^{-4}$.

  From the sketch, show that for $n \in \mathbb{Z}^+$, $\sum_{i=n+1}^{\infty} \dfrac{1}{i^4} < \int_n^{\infty} \dfrac{dx}{x^4} < \sum_{i=n}^{\infty} \dfrac{1}{i^4}$.

  b Let $S = \sum_{i=1}^{\infty} \dfrac{1}{i^4}$. Using the result from part **a**, show that for $n \geq 2$,

  $\sum_{i=1}^{n-1} \dfrac{1}{i^4} + \dfrac{1}{3n^3} \leq S \leq \sum_{i=1}^{n} \dfrac{1}{i^4} + \dfrac{1}{3n^3}$.

  c i Show that for $n = 8$, the value of S can be deduced to 3 d.p., and find this value.

   ii The exact value of S is known to be $\dfrac{\pi^4}{N}$, $N \in \mathbb{Z}^+$. Find the value of N.

10 Let $S_n = \sum_{i=1}^{n} \dfrac{1}{i}$.

  a Show that for $n \geq 2$, $S_{2n} > S_n + \dfrac{1}{2}$.

  b Deduce that $S_{2^{m+1}} > S_2 + \dfrac{m}{2}$.

  c Hence show that the sequence $\{S_n\}$ is divergent.

11 Show that the series $\sum_{n=1}^{\infty} (-1)^{n+1} \dfrac{n}{2n^2 + 1}$ is convergent, and hence show that its sum to infinity is less than 0.25.

12 a Prove that the series $\sum_{n=1}^{\infty} \dfrac{(-1)^{n-1}}{(2n-1)!}$ converges.

  b By finding the 4th partial sum, approximate the sum of the series to 6 d.p., and hence determine the upper bound of the error.

Chapter 4 summary

Definition: If the sequence of partial sums has a limit L as $n \to \infty$, then the series **converges** to the sum L, i.e. $u_1 + u_2 + u_3 + \ldots + u_n + \ldots = \sum_{k=1}^{\infty} u_k = L$.
Otherwise the series **diverges**.

The nth term test for divergence: If the sequence $\{u_n\}$ does not converge to 0, then the infinite series $\sum u_n$ diverges.

Definite Integral as the limit of a Riemann sum: Given that

f is a continuous function defined on the interval $[a, b]$

the interval $[a, b]$ is divided into n equal subintervals of width $\Delta x = \dfrac{b-a}{n}$

$x_i = a + i\Delta x$ is the right endpoint of the subinterval i

then the definite integral of f from a to b is $\displaystyle\int_a^b f(x)dx = \lim_{n \to \infty} \sum_{i=1}^{n} f(x_i)\Delta x$.

Fundamental Theorem of Calculus (FTC): If f is continuous in $[a, b]$, then

$\displaystyle\int_a^b f(x)dx = F(a) - F(b)$, where F is any function such that $F'(x) = f(x)$ for all x in $[a, b]$.

The FTC can also be expressed as $\dfrac{d}{dx}\left[\displaystyle\int_a^x f(t)dt\right] = f(x)$.

Definition: An improper integral $\displaystyle\int_a^b f(x)dx$ has infinity as one or both of its endpoints, or contains a finite number of infinite discontinuities in the interval $[a, b]$.

Improper Integral Notation

If $f(x)$ is continuous on $[a, \infty)$, then $\displaystyle\int_a^{\infty} f(x)dx = \lim_{b \to \infty} \int_a^b f(x)dx$.

Convergence Tests

Integral test: Let $f(x)$ be a continuous, positive, and decreasing function. Then the series $f(1) + f(2) + \ldots + f(n) + \ldots$ converges if the improper integral $\displaystyle\int_1^{\infty} f(x)dx$ converges, and diverges if the integral diverges. (The proof of this theorem uses concepts beyond the scope of the course.)

Theorem: If the alternating series $\displaystyle\sum_{n=1}^{\infty}(-1)^{n+1}u_n$ satisfies the conditions to apply Leibniz's theorem, then the truncation error for the nth partial sum is less than u_{n+1}, and has the same sign as the unused term. The sum of the series S_{∞} is therefore $S_n \pm$ the truncation error, i.e., $|S_{\infty} - S_n| = |R_n| \leq u_{n+1}$, where R_n is the truncation error, or remainder.

5 Everything polynomic

CHAPTER OBJECTIVES:

9.2 Power series: radius of convergence and interval of convergence; Determination of the radius of convergence by the ratio test.

9.6 Taylor Polynomials; the Lagrange form of the error term; Maclaurin series for e^x, $\sin(x)$, $\cos(x)$, $\ln(1+x)$, $(1+x)^p$, $p \in \mathbb{Q}$; Use of substitution, products, integration, and differentiation to obtain other series; Taylor series developed from differential equations;

9.7 Using L'Hopital rule or the Taylor series.

Before you start

You should know how to:

1 Manipulate infinite geometric series, e.g. Find the values of x for which the sum $1 + 2x + 4x^2 + 8x^3 + \ldots + (2x)^n + \ldots$ exists, and express the sum in terms of x. This is a geometric series for $|2x| < 1$ or $|x| < \frac{1}{2}$, hence $a = 1$ and $r = 2x$, so $S_\infty = \frac{1}{1-2x}$.

2 Differentiate all functions from the core syllabus using the different differentiation techniques you have learned, e.g. find the derivatives of:
   a $y = e^{x^2}$; $\quad y' = 2xe^{x^2}$
   b $y = \arcsin 3x$; $\quad \dfrac{dy}{dx} = \dfrac{3}{\sqrt{1-9x^2}}$
   c $y = \ln(2x+1)$; $\quad y' = \dfrac{2}{(2x+1)}$.

3 Integrate all functions from the core syllabus, e.g.
   a $\displaystyle\int \dfrac{dx}{1-x} = -\ln(1-x) + c$
   b $\displaystyle\int \dfrac{dx}{9+x^2} = \dfrac{1}{3}\arctan\left(\dfrac{x}{3}\right) + c$
   c $\displaystyle\int (e^{5x} - \sin x)\,dx = \dfrac{e^{5x}}{5} + \cos x + c$

Skills check:

1 **a** Find the values of x for which the series $\dfrac{1}{2} - \dfrac{x}{4} + \dfrac{x^2}{8} - \dfrac{x}{16} + \ldots + (-1)^n \dfrac{x^n}{2^{n+1}} + \ldots$ converges, and state the sum in terms of x.
   b Identify the values of x such that $\dfrac{2}{2-3x}$ is the sum of an infinite geometric series, and write out the series.

2 Find the derivatives of:
   a $y = \arctan 2x$
   b $y = \ln(1 - 3x)$
   c $y = e^{\sin(x)}$

3 Integrate with respect to x:
   a $y = \cos(x) + e^{-\frac{x}{2}}$
   b $y = \dfrac{3}{(4+x^2)}$
   c $y = \dfrac{1}{(1-4x)}$

From Finite to 'Infinite' Polynomials

In the early 19th century, the Norwegian mathematician Niels Henrik Abel completed Cauchy's work by clarifying many aspects of the convergence and divergence of so-called power series. Abel particularly despised using any divergent series in calculations, claiming that 'the divergent series are the invention of the devil'.

In the mid and late 19th century, Karl Weierstrass laid the foundations of modern analysis as it is studied today. He clarified many misconceptions by rigorously developing the notion of *uniform convergence* for series whose terms are functions, e.g. for power series; where every term is a function of x.

In this chapter you will learn how to represent functions as series, and represent series as functions. In a sense, much of the mathematics done up until now could have been done more simply by using 'infinite polynomials' or power series, as you will see during the development of this chapter.

> Karl Weierstrass (1815–97) is known as the 'father of modern analysis'. Weierstrass was concerned with the soundness of calculus, and sought to provide theorems which could be proven with sufficient rigour.

5.1 Representing Functions by Power Series 1

Consider the series $1 + x + x^2 + x^3 + \ldots + x^n + \ldots$ whose partial sums are all polynomials. You have seen this before, in the form of an infinite geometric series. We know that if $|x| < 1$, then the series has the sum $\frac{1}{1-x}$. *With some caution*, we can therefore equate the two expressions: $1 + x^2 + x^3 + \ldots + x^n + \ldots = \frac{1}{1-x}$. The right hand side defines a function whose domain is the set of real numbers, except for $x = 1$. The left hand side defines a function whose domain is the interval of convergence $|x| < 1$ or $x \in]-1, 1[$. The equality therefore only holds in the domain that is the interval of convergence. On this domain, the infinite polynomial series represents the function $f(x) = \frac{1}{1-x}$.

The graph on the right shows some of the partial sums of the polynomial series, together with the function representing the sum of the infinite polynomial series. As you would expect, the more partial sums of the polynomial series we graph, the stronger the convergence between the series and the function. However, this happens only in the interval of convergence, $]-1, 1[$. Outside of this interval, the convergence breaks down.

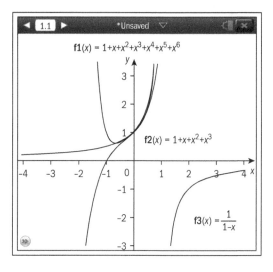

We need to mention here that the expression $\sum_{n=0}^{\infty} x^n$ is similar to a polynomial, but polynomial functions have finite degrees and we do not refer to the convergence of a polynomial function. The expression therefore is not just a polynomial, but instead we call it a *power series*.

Definition: An expression of the form 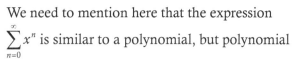 $= c_0 + c_1 x + c_2 x^2 + \ldots + c_n x^n + \ldots$,

where $c_n \in \mathbb{R}$ for all n, is a **power series centered at $x = 0$.**

The set of points for which the series converges is the **interval of convergence**, and this set consists of the interval $]-R, R[$ where R is the **radius of convergence**.

The geometric series $1 + x^2 + x^3 + \ldots + x^n + \ldots$ is therefore a power series centered at $x = 0$. It converges on the interval $]-1, 1[$; whose center is $x = 0$, and radius $R = 1$.

> ❓ The quantity R is called the radius of convergence because, in the case of power series with complex numbers, the interval of convergence forms a circle with radius R.

Example 1

a Find a power series that represents the function $y = \dfrac{1}{1+x}$ on the interval $]-1, 1[$.	
b Find a power series that represents the function $y = \dfrac{1}{1-2x}$ on the interval $\left]-\dfrac{1}{2}, \dfrac{1}{2}\right[$.	

a In this interval $\dfrac{1}{1+x}$ is the sum of the infinite geometric series when the common ratio is $-x$, and $\|x\| < 1$.	*Recognize the function as an infinite geometric series when $\|r\| < 1$.*
So $\dfrac{1}{1+x} = 1 - x + x^2 - x^3 + \ldots + (-x)^n + \ldots$	*Write the function as a power series.*
with $I = \,]-1, 1[\,;\ R = 1$	*I is the interval of convergence*
b $\dfrac{1}{1-2x}$ is the sum of an infinite series when the common ratio, $2x$, is such that $\|2x\| < 1$, or $\|x\| < \dfrac{1}{2};\ R = \dfrac{1}{2}$	*Recognize the function as an infinite geometric series when $\|r\| < 1$.*
So $\dfrac{1}{1-2x} = 1 + 2x + (2x)^2 + (2x)^3 + \ldots + (2x)^n + \ldots$ $= 1 + 2x + 4x^2 + 8x^3 + \ldots + 2^n x^n + \ldots$	*Write the function as a power series.*

Are there power series whose interval of convergence is not centered at $x = 0$? Consider the function $f(x) = \dfrac{1}{1-(x-1)}$, and let us attempt to find a power series to represent this function (notice that the function is the same as $f(x) = \dfrac{1}{2-x}$).

In this case, $r = (x-1)$, and so the interval of convergence is $|r| < 1$, or $-1 < (x-1) < 1$, i.e. $0 < x < 2$. The power series representing the function $\dfrac{1}{1-(x-1)}$ is therefore

$$1 + (x-1) + (x-1)^2 + (x-1)^3 + \ldots + (x-1)^n + \ldots = \sum_{n=0}^{\infty}(x-1)^n$$

and is centered at $x = 1$. The radius of convergence is $]1-R, 1+R[$ or $R = 1$.

Definition: An expression of the form

$$\sum_{n=0}^{\infty} c_n(x-a)^n = c_0 + c_1(x-a) + c_2(x-a)^2 + \ldots + c_n(x-a)^n + \ldots,$$

where $c_n, a \in \mathbb{R}$, is a power series centered at $x = a$. The set of points for which the series converges is the **interval of convergence**, and this set consists of the interval $]a - R, a + R[$ where R is the **radius of convergence**.

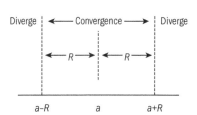

Example 2

a Find a power series to represent $f(x) = \dfrac{1}{x}$, stating the interval and radius of convergence.

b Find a power series to represent $f(x) = \dfrac{1}{2+x}$, and state the interval and radius of convergence.

c Find a power series to represent $f(x) = \dfrac{x^2}{2+x}$, and state the interval and radius of convergence.

---

a $\dfrac{1}{x} = \dfrac{1}{1+(x-1)}$ | Change the function to the form of an infinite geometric series.

$\dfrac{1}{1+(x-1)} = \sum_{n=0}^{\infty} (-1)^n (x-1)^n$ | Write the series in general form and show the term expansion.

$= 1 - (x-1) + (x-1)^2 - (x-1)^3 + \ldots + (-1)^n (x-1)^n + \ldots$

Interval of convergence is $|-(x-1)| < 1$ or $I =]0, 2[$, and $R = 1$. | Identify the domain (or interval of convergence) I, and radius of convergence, R.

b $\dfrac{1}{2+x} = \dfrac{1}{2\left(1+\dfrac{x}{2}\right)} = \dfrac{1}{2} \cdot \left(\dfrac{1}{1-\left(-\dfrac{x}{2}\right)}\right)$ | Change the function to the form of an infinite geometric series.

$= \dfrac{1}{2} \sum_{n=0}^{\infty} \left(-\dfrac{x}{2}\right)^n ; \left|-\dfrac{x}{2}\right| < 1$ | Write the series in general form.

$f(x) = \dfrac{1}{2} \sum_{n=0}^{\infty} \left(-\dfrac{x}{2}\right)^n = \sum_{n=0}^{\infty} \dfrac{1}{2} \cdot \dfrac{(-1)^n}{2^n} x^n = \sum_{n=0}^{\infty} \dfrac{(-1)^n}{2^{n+1}} x^n$ | Write out $f(x)$.

$= \dfrac{1}{2} - \dfrac{x}{4} + \dfrac{x^2}{8} - \dfrac{x^3}{16} + \ldots + \dfrac{(-1)^n x^n}{2^{n+1}} + \ldots$

$\left|-\dfrac{x}{2}\right| < 1 \Rightarrow |x| < 2, I =]-2, 2[\; ; R = 2$ | Find the interval and radius of convergence.

c From part (b), $f(x) = x^2 \sum_{n=0}^{\infty} \dfrac{(-1)^n}{2^{n+1}} x^n ; |x| < 2$ | Change the function to the form of an infinite geometric series.

$= \sum_{n=0}^{\infty} \dfrac{(-1)^n}{2^{n+1}} x^{n+2} ; |x| < 2$ | Write out $f(x)$ in general form.

$= \dfrac{x^2}{2} - \dfrac{x^3}{4} + \ldots + \dfrac{(-1)^n}{2^{n+1}} x^{n+2} + \ldots$ | Write out some terms.

$I =]-2, 2[, R = 2.$ | State the interval and radius of convergence.

Exercise 5A

Find a power series to represent the following functions and state the interval and radius of convergence.

a $f(x) = \dfrac{1}{1-4x}$

b $f(x) = \dfrac{1}{1+5x}$

c $f(x) = \dfrac{1}{x-4}$

d $f(x) = \dfrac{1}{3x}$

e $f(x) = \dfrac{x}{1+x}$

f $f(x) = \dfrac{2}{1+4x}$

5.2 Representing Power Series as Functions

A power series can be written as a function, i.e. $f(x) = \sum_{n=0}^{\infty} c_n (x-a)^n$, where the domain of f is the set of all x for which the power series converges. The main task in working with power series is to determine this domain.

We can see from the above definition that every power series converges at its center a, since

$$f(a) = \sum_{n=0}^{\infty} c_n (a-a)^n = c_0 + 0 + 0 \ldots + 0 + \ldots = c_0.$$

Hence, a always lies in the domain of f.

Sometimes the domain of f consists only of $x = a$; sometimes it is an interval centered at a, as we have seen in examples 1 and 2; and other times the domain is all x (as we show below):

Consider the power series $f(x) = \sum_{n=0}^{\infty} \dfrac{x^n}{n^n} = 1 + x + \dfrac{x^2}{2^2} + \dfrac{x^3}{3^3} + \ldots + \dfrac{x^n}{n^n} + \ldots$

We can observe that for any fixed x, n will eventually be greater than $2|x|$, which means that $\left|\dfrac{x}{n}\right| < \dfrac{1}{2}$. Hence, $\left|\dfrac{x^n}{n^n}\right| < \dfrac{1}{2^n}$, for sufficiently large n.

The series therefore converges absolutely by comparison with the geometric series, and the domain or interval of convergence is all real x.

> ❓ You will have probably noticed that to achieve 1 as the first term in the power series above, we have defined $0^0 = 1$. Mathematicians have been discussing the question of the meaning and value of 0^0 for centuries. Euler argued vigorously for defining 0^0 as 1. By and large, defining 0^0 as 1 is a useful convention that allows mathematicians to extend definitions in other areas of mathematics that would otherwise need to treat 0^0 as a special case.

From the above examples we can now formulate the following theorem.

Theorem 1: For a power series $f(x) = \sum_{n=0}^{\infty} c_n (x-a)$ centered at a, either:
- The series converges only at a; or
- The series converges for all x; or
- The series converges for $|x - a| < R$, where $R > 0$.

In examples 1 and 2 we have found the interval of convergence of a power series that is geometric. To determine where other power series converge, if at all, the Ratio Test done in chapter 4 is very useful.

Consider the power series $f(x) = \sum_{n=1}^{\infty} n! x^n = x + 2x^2 + 6x^3 + \ldots + n! x^n + \ldots$

We can immediately see that when $x = 0$, $f(0) = 0$, hence the power series converges at $x = 0$, its center. We can now test for absolute convergence using the ratio test.

$$\lim_{n \to \infty} \left| \frac{u_{n+1}}{u_n} \right| = \lim_{n \to \infty} \left| \frac{(n+1)! x^{n+1}}{n! x^n} \right| = |x| \lim_{n \to \infty} (n+1) = \infty. \text{ Since } L > 1, \text{ the series}$$

diverges for all values of x, except $x = 0$. The series converges only when $x = 0$.

Ratio Test Applied to Power Series: For a power series

$f(x) = \sum_{n=0}^{\infty} c_n (x-a)^n$, if

- $\lim_{n \to \infty} \left| \frac{a_{n+1}}{a_n} \right| = L$ and $L \neq 0$, then $R = \dfrac{1}{L}$

- $\lim_{n \to \infty} \left| \frac{a_{n+1}}{a_n} \right| = 0$, then the radius of convergence is infinite

- $\lim_{n \to \infty} \left| \frac{a_{n+1}}{a_n} \right| = \infty$, then $R = 0$.

Up until this point we have not considered the endpoints of the interval in our discussions. The theorems and tests to date only give us the open interval of convergence, i.e. they do not tell us whether a power series converges when x lies at either of the endpoints of the interval. These endpoints must be tested separately.

For example, consider the power series $f(x) = \sum_{n=1}^{\infty} \dfrac{x^n}{n}$.

Using the ratio test for power series, we obtain

$$\lim_{n \to \infty} \left| \frac{u_{n+1}}{u_n} \right| = \lim_{n \to \infty} \left| \frac{\frac{x^{n+1}}{n+1}}{\frac{x^n}{n}} \right| = \lim_{n \to \infty} \left[|x| \frac{n}{n+1} \right] = |x| \lim_{n \to \infty} \frac{n}{n+1} = |x|.$$

For the series to converge, the limit must be less than 1.
Hence, $|x| < 1$ is the interval of convergence, and $R = 1$.

We will now test the end points of the interval:

When $x = 1$, $\sum_{n=1}^{\infty} \frac{1}{n}$ is the harmonic series, and therefore diverges.

When $x = -1$, $\sum_{n=1}^{\infty} \frac{(-1)^n}{n}$ is the alternating harmonic series, which we know converges.

Therefore, the interval of convergence for the given series is $[-1, 1[$.

Example 3

Find the radius and interval of convergence, including the endpoints, for the following series

a $f(x) = \sum_{n=1}^{\infty} \frac{x^n}{n^2}$.

b $f(x) = \sum_{n=0}^{\infty} \frac{(-2)^n x^n}{\sqrt{n+1}}$

---

a $\lim_{n \to \infty} \left| \frac{\frac{x^{n+1}}{(n+1)^2}}{\frac{x^n}{n^2}} \right| = |x| \lim_{n \to \infty} \frac{n^2}{(n+1)^2} = |x|$

Use the ratio test

The series converges for $x \in [-1, 1]$, and $R = 1$

Find interval of convergence using the condition $L < 1$, and from the earlier result, the radius of convergence is $R = 1/mod(x)$. Test convergence at endpoints.

At $x = 1$, $\sum_{n=1}^{\infty} \frac{1}{n^2}$ converges by the p-series test;

at $x = -1$, $\sum_{n=1}^{\infty} \frac{(-1)^n}{n^2}$ converges by the alternating series test. Hence, $I =]-1, 1[$.

b $\lim_{n \to \infty} \left| \frac{\frac{(-2)^{n+1} x^{n+1}}{\sqrt{n+2}}}{\frac{(-2)^n x^n}{\sqrt{n+1}}} \right| = 2|x| \lim_{n \to \infty} \frac{\sqrt{n+1}}{\sqrt{n+2}} = 2|x|$

Use ratio test.

The series converges in the interval $2|x| < 1$, or $I = \left]-\frac{1}{2}, \frac{1}{2}\right[$ and $R = \frac{1}{2}$.

Find interval of convergence using the condition $L < 1$, and state R.

At $x = \frac{1}{2}$,

$\sum_{n=0}^{\infty} \frac{(-2)^n x^n}{\sqrt{n+1}} = \sum_{n=0}^{\infty} \frac{(-2)^n \left(\frac{1}{2}\right)^n}{\sqrt{n+1}} = \sum_{n=0}^{\infty} \frac{(-1)^n}{\sqrt{n+1}}$.

Determine convergence at the endpoints.

Converges by the alternating series test;

at $x = -\frac{1}{2}$, $\sum_{n=0}^{\infty} \frac{(-2)^n x^n}{\sqrt{n+1}} = \sum_{n=0}^{\infty} \frac{1}{\sqrt{n+1}}$, which diverges by the p-series test. Hence, $I = \left]-\frac{1}{2}, \frac{1}{2}\right]$.

Exercise 5B

1 Find the radius and interval of convergence, including testing the endpoints, of the following power series.

a $\sum_{n=1}^{\infty}(-1)^n \dfrac{x^n}{n}$

b $\sum_{n=0}^{\infty}\dfrac{x^n}{n!}$

c $\sum_{n=0}^{\infty}\dfrac{2^n x^n}{(n+1)^2}$

d $\sum_{n=1}^{\infty}(-1)^n \dfrac{(2x+3)^n}{n \ln n}$

e $\sum_{n=0}^{\infty}(-1)^{n+1}\dfrac{(x-1)^{n+1}}{n+1}$

f $\sum_{n=0}^{\infty}\dfrac{x^{2n+1}}{(2n+1)!}$

g $\sum_{n=1}^{\infty}\dfrac{x^n}{n\sqrt{n}3^n}$

h $\sum_{n=0}^{\infty}\dfrac{\sqrt{n}x^n}{3^n}$

i $\sum_{n=1}^{\infty}\dfrac{\sin^n x}{n^2}$

j $\sum_{n=1}^{\infty}\dfrac{e^{nx}}{n}$

k $\sum_{n=1}^{\infty}\dfrac{(x+1)^n}{3^n}$

2 Find the value of k in the following power series with the given interval of convergence:

a $\sum_{n=1}^{\infty}\dfrac{(kx)^n}{n^2}; \left[-\dfrac{1}{2}, \dfrac{1}{2}\right]$

b $\sum_{n=2}^{\infty}\dfrac{(x-k)^{2n}}{n^4}; [2, 4]$

3 Find h and k if the interval of convergence of the series $\sum_{n=1}^{\infty}\dfrac{(x-h)^n}{nk^n}$ is $]-1, 7[$.

5.3 Representing Functions by Power Series 2

In the beginning of this chapter we saw how to represent functions using an infinite geometric series. Do we have other methods of representing functions by power series? Since the partial sums that converge to the power series are polynomials, we can use calculus to obtain further results, as the examples will show.

We know that the function $y = \dfrac{1}{1-x}$ is represented by a power series on the interval $|x| < 1$, and is centered at $x = 0$. Can we use this fact to find a power series to represent the function $y = \dfrac{1}{(1-x)^2}$ on the same interval of convergence?

Since $\dfrac{1}{1-x} = 1 + x + x^2 + x^3 + \ldots x^n + \ldots$, we can differentiate the left and right hand sides $\dfrac{d}{dx}\left(\dfrac{1}{1-x}\right) = \dfrac{d}{dx}(1 + x + x^2 + x^3 + \ldots + x^n \ldots)$ and obtain $\dfrac{1}{(1-x)^2} = 1 + 2x + 3x^2 + 4x^3 + \ldots + nx^{n-1} + \ldots$

Since the interval of convergence of the original series is $|x| < 1$, it would seem that the differentiated series would have the same interval of convergence, except perhaps the end-points. Graphing the function and some of the partial sums of the series, we obtain

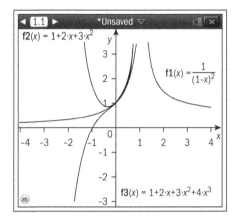

It seems as if the interval of convergence of the differentiated series and the original series are the same, except for possibly the end-points, and indeed they are. The proof however requires some concepts beyond the level of the course, so we will state the theorem without its proof.

Theorem 2: If $\sum_{n=0}^{\infty} c_n (x-a)^n = c_0 + c_1(x-a) + c_2(x-a)^2 + \ldots + c_n(x-a)^n + \ldots$ converges in the interval $|x-a| < R$ where $R > 0$, then the series

$f'(x) = \sum_{c=1}^{\infty} nc_n (x-a)^{n-1}$, obtained by differentiating the original series

term by term, also converges in the interval $|x-a| < R$. Furthermore, if the series for f converges for all x, then so does the series for f'.

Having found a power series by differentiating term by term, the next logical question is: Can we also find a power series by integrating term by term?

For example, can we find a power series to represent the function $\ln(1 + x)$?

Recognizing that $\int \frac{1}{1+x} dx = \ln(1+x) + c$, we can start with

$\frac{1}{1+x} = 1 - x + x^2 - x^3 + \ldots + (-1)^n x^n + \ldots$ in the interval $|-x| < 1$, or $x \in \,]-1, 1[$.

Integrating both sides, and using a dummy variable t, we obtain

$$\int_0^x \frac{1}{1+t} dt = \int_0^x \left(1 - t + t^2 - t^3 + \ldots + (-1)^n t^n + \ldots\right) dt$$

Integrating both sides

$$\left[\ln(1+t)\right]_0^x = \left[t - \frac{t^2}{2} + \frac{t^3}{3} + \ldots + (-1)^n \frac{t^{n+1}}{n+1} + \ldots\right]_0^x$$

and evaluating at the lower and upper bounds of the integral

$$\ln(1+x) = x - \frac{x^2}{2} + \frac{x^3}{3} + \ldots + (-1)^n \frac{x^{n+1}}{n+1} + \ldots$$

Again, we can see from the GDC that it appears the interval of convergence is the same for both the original series and the new series. For the power series, only the 4th partial sum is graphed. The higher the partial sums that are graphed, the stronger the convergence of power series to the function itself, on the interval.

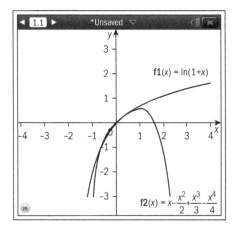

We can now state the following theorem without its proof, since the proof involves concepts beyond the level of the course.

Theorem 3: If $f(x) = \sum_{n=0}^{\infty} c_n (x-a)^n$ converges for $|x-a| < R$ and $R > 0$, then the series $\sum_{n=0}^{\infty} c_n \frac{(x-a)^{n+1}}{n+1}$ obtained by integrating f term by term also converges on the same interval and represents $\int_a^x f(t) \, dt$ on that interval. Furthermore, if the series for f converges for all x, then the series for the integral of f also converges for all x.

We know that the function $f(x) = \frac{1}{1+x}$ can be written as a geometric series for $|x| < 1$, and hence we know its interval of convergence is $]-1, 1[$. This series clearly diverges at the endpoints of the interval, $x = \pm 1$. The interval of convergence therefore will be the same for $f(x) = \ln(1+x)$, except possibly at the endpoints. When $x = -1$, we obtain $\ln(0)$, which is undefined. At $x = 1$, we obtain

$$\ln 2 = 1 - \frac{1}{2} + \frac{1}{3} - \frac{1}{4} + \ldots + \frac{(-1)^n}{n+1} + \ldots = \sum_{n=0}^{\infty}(-1)^n \frac{1}{n+1}.$$

We recognize this as the alternating harmonic series, which we know converges conditionally.

Hence, the interval of convergence for $f(x) = \ln(x+1)$ is $]-1, 1]$.

Example 4

a Find a power series for $f(x) = \arctan x$.
b Find a power series for $f(x) = x\arctan(x)$ centered at $x = 0$.

a $\arctan x + c = \int \frac{dx}{1+x^2}$; $\frac{1}{1+x^2} = 1 - x^2 + x^4 - x^6 + \ldots + (-1)^n (x^2)^n + \ldots$; I is $]-1, 1[$, $R = 1$.	Identify a series whose convergence you can determine.
$\int_0^x \frac{1}{1+t^2} dt = \int_0^x \left(1 - t^2 + t^4 - t^6 + \ldots + (-1)^n t^{2n}\right) + \ldots$	Integrate both sides.
$\arctan x = x - \frac{x^3}{3} + \frac{x^5}{5} - \frac{x^7}{7} + \ldots + (-1)^n \frac{x^{2n+1}}{2n+1} + \ldots$	Write out the power series.
$\arctan x = \sum_{n=0}^{\infty}(-1)^n \frac{x^{2n+1}}{2n+1}$	
b From part **a**, $\arctan x = \sum_{n=0}^{\infty}(-1)^n \frac{x^{2n+1}}{2n+1}$ I is $]-1, 1[$, $R = 1$	Find a power series for $\arctan(x)$.
$x\arctan(x) = x\sum_{n=0}^{\infty}(-1)^n \frac{x^{2n+1}}{2n+1} = \sum_{n=0}^{\infty}(-1)^n \frac{x^{2n+2}}{2n+1}$ I is $]-1, 1[$, $R = 1$.	Multiply the power series by x.

We have been able to apply calculus operations to generate power series that represent functions. We will now consider the arithmetic operations with regard to power series.

Theorem 4: Given the power series represented by the functions f and g, where $f(x) = \sum_{n=0}^{\infty} c_n(x-a)^n$ and $g(x) = \sum_{n=0}^{\infty} d_n(x-a)^n$ the following properties hold:

- $f(kx) = \sum_{n=0}^{\infty} c_n k^n (x-a)^n$

- $f(x) \pm g(x) = \sum_{n=0}^{\infty} (c_n \pm d_n)(x-a)^n$

- $f(x) \cdot g(x) = \left(\sum_{n=0}^{\infty} c_n(x-a)^n \right) \left(\sum_{n=0}^{\infty} d_n(x-a)^n \right)$

Chapter 5

Let the radius of convergence of f be R_f and g be R_g. Then

- Radius of convergence of the sum or difference is:
  - $\min\{R_f, R_g\}$ when $R_f \neq R_g$, i.e. the minimum of the two radii; or
  - greater than or equal to the radii when $R_f = R_g$.
- Radius of convergence of the product is $\min\{R_f, R_g\}$.

Consider the power series defined by $f(x) = \sum_{n=0}^{\infty} x^n$ and $g(x) = \sum_{n=0}^{\infty} \left(\frac{x}{2}\right)^n$.

Then $f(x) + g(x) = \sum_{n=0}^{\infty} x^n + \sum_{n=0}^{\infty} \left(\frac{x}{2}\right)^n = \sum_{n=0}^{\infty} \left(1 + \frac{1}{2^n}\right) x^n$.

Since both f and g are geometric series, it is easy to establish that the interval of convergence of f is $]-1, 1[$ with $R = 1$, and the interval of convergence of g is $I =]-2, 2[$ with $R = 2$.

The interval of convergence of the sum and the product of power series is the intersection of the interval of convergence of f and g, i.e. $I =]-1, 1[$, and $R = \min\{1, 2\} = 1$.

Example 5

Show that $\dfrac{7x-2}{x^2-x-2}$, $x \neq -1, 2$, can be written as $\dfrac{3}{x+1} + \dfrac{4}{x-2}$ and hence find a power series centered at $x = 0$ for $f(x) = \dfrac{7x-2}{x^2-x-2}$. Determine its interval and radius of convergence.

$\dfrac{3}{x+1} + \dfrac{4}{x-2} = \dfrac{3(x-2)+4(x+1)}{(x+1)(x-2)} = \dfrac{3x-6+4x+4}{x^2-x-2} = \dfrac{7x-2}{x^2-x-2}$	Add the algebraic fractions.
$\dfrac{3}{x+1} = 3\left(\dfrac{1}{x+1}\right) = 3\sum_{n=0}^{\infty} (-1)^n x^n$ $I =]-1, 1[, R = 1$	Recognize the function as an infinite geometric series with $\|r\| < 1$.
$\dfrac{4}{x-2} = 4\left(\dfrac{1}{x-2}\right) = 4\left(\dfrac{1}{-2\left(1-\dfrac{x}{2}\right)}\right)$ $= -\dfrac{4}{2} \sum_{n=0}^{\infty} \left(\dfrac{x}{2}\right)^n = -2 \sum_{n=0}^{\infty} \left(\dfrac{x}{2}\right)^n = \sum_{n=0}^{\infty} -2 \left(\dfrac{x}{2}\right)^n$; $I =]-2, 2[; R = 2$	Change the function to the form of an infinite geometric series, and write the series in general form.
$\dfrac{7x-2}{x^2-x-1} = \sum_{n=0}^{\infty} 3(-1)^n x^n + \sum_{n=0}^{\infty} -2\left(\dfrac{x}{2}\right)^n$ $= \sum_{n=0}^{\infty} \left(3(-1)^n - \dfrac{2}{2^n}\right) x^n = \sum_{n=0}^{\infty} \left(\dfrac{3(-1)^n 2^{n-1} - 1}{2^{n-1}}\right) x^n$	Add the two series using arithmetic properties of power series.
$]-2, 2[\cap]-1, 1[=]-1, 1[, I =]-1, 1[; R = 1$	Find the intersection of convergence intervals for the two series and R.

Exercise 5C

Find a power series for the following functions. Determine the interval and radius of convergence. Unless otherwise stated, the center is $x = 0$.

1 $f(x) = \dfrac{2}{1-x^2}$

2 $f(x) = -\dfrac{1}{(x+1)^2}$

3 $f(x) = \dfrac{1}{4x^2+1}$

4 $f(x) = \dfrac{\sqrt{1+x}}{\sqrt{1-x}}$

5 $f(x) = \ln(x)$, center $x = 1$

6 The radius of convergence of $\sum_{n=0}^{\infty} a_n x^n$ is 1, and the radius of convergence of $\sum_{n=0}^{\infty} b_n x^n$ is 2. State the radius of convergence of $\sum_{n=0}^{\infty} (a_n x^n + b_n x^n)$.

7 Find the radius of convergence and the interval of convergence of the series $\sum_{n=1}^{\infty} \dfrac{x^n}{n^2 2^n}$ and $\sum_{n=1}^{\infty} \dfrac{nx^{n-1}}{n^2 2^n}$, and comment on your result.

8 Given that $f(x) = \dfrac{3x}{x^2+x-2}$, $g(x) = \dfrac{2}{x+2}$, and $h(x) = \dfrac{1}{x-1}$, show that $f(x) = g(x) + h(x)$, and find a power series for $f(x)$ centered at $x = 0$, and determine the interval of convergence.

5.4 Taylor Polynomials

In the previous sections we have looked at ways of representing functions by power series, and representing power series by functions. We will continue now our use of calculus to enable us to construct power series.

Investigation 1

Construct a polynomial $P(x) = a_0 + a_1 x + a_2 x^2 + a_3 x^3 + a_4 x^4 + a_5 x^5$ with the following conditions at $x = 0$:

$P(0) = 1$; $P'(0) = 2$; $P''(0) = 3$; $P'''(0) = 4$ and $P^{(4)}(0) = 5$; $P^{(5)}(0) = 6$.

Express the coefficients of x^n in terms of $P^{(n)}(x)$ and n.

The investigation above shows how we can construct a polynomial by knowing its behavior at one single point! There is nothing special about the first five derivatives of the polynomial. We could continue and construct the polynomial for $P^{(n)}(0)$.

We will use the plan in investigation 1 to construct a polynomial function, and observe the behavior of the function at $x = 0$.

Example 6

Construct a polynomial function $P(x) = a_0 + a_1x + a_2x^2 + a_3x^3 + a_4x^4 + a_5x^5$ whose behavior at $x = 0$ is similar to the behavior of $f(x) = \ln(1 + x)$ at $x = 0$ through its first four derivatives.

$P(0) = \ln(1 + 0) = 0;\ a_0 = 0$	Evaluate $f(x)$ at the given point and find a_n for $n = 0, 1, 2, 3, 4$.
$P'(x) = \dfrac{1}{1+x};\ P'(0) = 1;\ a_1 = 1$	
$P''(x) = -\dfrac{1}{(1+x)^2};\ P''(0) = -1;\ a_2 = -\dfrac{1}{2}$	
$P'''(x) = \dfrac{2}{(1+x)^3};\ P'''(0) = 2;\ a_3 = \dfrac{2}{6}$	
$P^{(4)}(x) = -\dfrac{6}{(1+x)^4};\ P^{(4)}(0) = -6;\ a_4 = -\dfrac{6}{24}$	
$P(x) = 0 + x - \dfrac{1}{2}x^2 + \dfrac{2}{6}x^3 - \dfrac{6}{24}x^4$	Write the first five terms of the power series for $f(x)$, and let this polynomial be called $P(x)$.
$P(x) = x - \dfrac{x^2}{2} + \dfrac{x^3}{3} - \dfrac{x^4}{4}$	Simplify
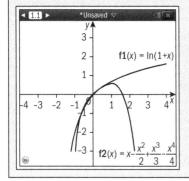	Graph the function and polynomial

Observe that we previously arrived at the same function by integrating $f(x) = \dfrac{1}{1+x}$ term by term. The higher the partial sums we graphed, the stronger the convergence on the interval centered at $x = 0$.

What we have just done in Example 3 is to construct the 4th order Taylor polynomial for the function $f(x) = \ln(1 + x)$ at $x = 0$.

Definition: If $f(x)$ has n derivatives at $x = a$, then the Taylor polynomial of degree n for $f(x)$ centered at a, $T_n(x)$, is the unique polynomial of degree n which satisfies

$$T_n(a) = f(a);\ T_n'(a) = f'(a);\ T_n''(a) = f''(a); \ldots;\ T^{(n)}(a) = f^{(n)}(a).$$

Furthermore,

$$T_n(x) = f(a) + \dfrac{f'(a)}{1!}(x-a) + \dfrac{f''(a)}{2!}(x-a)^2 + \ldots + \dfrac{f^{(n)}(a)}{n!}(x-a)^n = \sum_{k=0}^{n} \dfrac{f^{(k)}(a)}{k!}(x-a)^k.$$

> As early as the 14th century, the Indian mathematician Madhava of Sangamagrama had represented trigonometric functions as power series in order to solve problems in astronomy. It is thought by some that it was actually Madhava who invented the field of mathematical analysis. Later Sanskrit writings by the mathematician Jyesthadeva containing Madhava's work indicate that his work preceded that of western mathematicians by about 300 years. However, since his work has only been recently brought to light, the rich and intense development in mathematics is still attributable to the prolific writings of the mathematicians of the western world.

Example 7

Compute the 4th order Taylor polynomial for the function $f(x) = \sin x$ at $x = \frac{\pi}{4}$.

$f\left(\frac{\pi}{4}\right) = \frac{\sqrt{2}}{2}$	*Evaluate the function and its first four derivatives at the given point.*
$f'(x) = \cos(x);\ \cos\left(\frac{\pi}{4}\right) = \frac{\sqrt{2}}{2}$	
$f''(x) = -\sin x;\ f''\left(\frac{\pi}{4}\right) = -\frac{\sqrt{2}}{2}$	
$f'''(x) = -\cos x = f'''\left(\frac{\pi}{4}\right) = -\frac{\sqrt{2}}{2}$	
$f^{(4)}(x) = \sin x;\ f^{(4)}\left(\frac{\pi}{4}\right) = \frac{\sqrt{2}}{2}$	
$T_4(x) = \frac{\sqrt{2}}{2} + \frac{\sqrt{2}}{2}\left(x - \frac{\pi}{4}\right) - \frac{\sqrt{2}}{2 \cdot 2!}\left(x - \frac{\pi}{4}\right)^2 - \frac{\sqrt{2}}{2 \cdot 3!}\left(x - \frac{\pi}{4}\right)^3 + \frac{\sqrt{2}}{2 \cdot 4!}\left(x - \frac{\pi}{4}\right)^4$	*Write out the Taylor polynomial*

Graphing both the function and its 4th order Taylor polynomial, observe that the Taylor polynomial is a good approximation of the function at $x = \frac{\pi}{4}$.

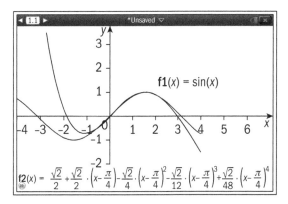

Note that the properties on page 141 for power series also hold true for Taylor polynomials. That is, the Taylor polynomial for the sum of two functions f and g is the sum of the Taylor polynomials for f and g, and the product of the Taylor polynomials for f and g is the Taylor polynomial of the product of f and g.

Hence, for example, if we want to find the 4th order Taylor polynomial of $f(x) = \sin(x) + \cos(x)$, we simply find the 4th order Taylor polynomial for $\sin(x)$ and the 4th order Taylor polynomial for $\cos(x)$, and add them. The same holds for finding the Taylor polynomial of, for example, $f(x) = \sin(x)\cos(x)$, i.e. find the Taylor polynomial for each function separately, and then find the product of the two polynomials.

Exercise 5D

1. Compute the 4th order Taylor polynomial, centered at $x = 0$, for the following functions.

   a $f(x) = \dfrac{1}{1-x}$ b $f(x) = x\sin(x)$ c $f(x) = \sqrt[3]{1+x}$

   d $f(x) = xe^x$

2. Construct the 3rd order Taylor polynomial, centered at $x = \pi$, for the following functions.

   a $f(x) = \dfrac{1}{x}$ b $f(x) = \dfrac{\sin x}{x}$ c $f(x) = x^2 \cos x$

3. Find the third order Taylor polynomial for $f(x) = 2x^3 - 3x^2 + 4x - 5$ at

   a $x = 0$ and   b $x = 1$.

4. Explain why the first order Taylor polynomial for $f(x)$ centered at $x = a$ is the equation of the tangent line to the graph of f at a.

5. Write the third order Taylor polynomial for f at $x = 1$, and use it to approximate $f(1.2)$, given that $f(x)$ is a function that has derivatives of all orders for all real numbers, and $f(1) = 4$, $f'(1) = -1$, $f''(1) = 3$, $f'''(1) = 2$.

5.5 Taylor and Maclaurin Series

We will begin this section by constructing the nth degree Taylor polynomial for the function $f(x) = e^x$ at $x = 0$.

Computing successive derivatives gives us the same function e^x. Evaluating the function and its derivatives at $x = 0$, we obtain the same result, $f^n(0) = 1$ for all n.

Hence, $T_n(x) = f(0) + f'(0)x + \dfrac{1}{2}f''(0)x^2 + \ldots + \dfrac{1}{n!}f^{(n)}(0)x^n$, or

$$T_n(x) = 1 + x + \dfrac{x^2}{2!} + \dfrac{x^3}{3!} + \ldots + \dfrac{x^n}{n!} = \sum_{k=0}^{n} \dfrac{x^k}{k!}.$$

Looking at both the graph and the table of values of the first seven terms of the Taylor polynomial, we can confirm the accuracy of the approximation.

Since the function $f(x) = e^x$ has derivatives of all orders at $x = 0$, we can continue obtaining and evaluating derivatives, thereby improving our approximation near $x = 0$ with each term we add, obtaining an infinite series.

Definition: If $f(x)$ has derivatives of all orders throughout an open interval I such that $a \in I$, then the Taylor Series generated by f at $x = a$ is

$$f(x) = f(a) + f'(a)(x-a) + \frac{f''(a)}{2!}(x-a)^2 + \frac{f'''(a)}{3!}(x-a)^3 + \dots$$

$$= \sum_{k=0}^{\infty} \frac{f^{(k)}(a)}{k!}(x-a)^k.$$

If $a = 0$, then the Taylor Series is also referred to as the **Maclaurin series** generated by f.

The partial sums of both the Taylor and Maclaurin Series have already been defined, and are $T_n(x) = \sum_{k=0}^{n} \frac{f^{(k)}(a)}{k!}(x-a)^k$.

If we define $c_n = \frac{f^{(n)}(a)}{n!}$, then you will notice that the definition of the Taylor series is the same as that of a power series, where

$$T_n(x) = \sum_{T_n(x)=0}^{\infty} c_n(x-a)^n = c_0 + c_1(x-a) + c_2(x-a)^2 + \dots + c_n(x-a)^n + \dots$$

> Interestingly enough, neither Brook Taylor (1685–1731) nor Colin Maclaurin (1698–1746) was the first to define the series that are named after them. The Scottish mathematician James Gregory (1638–1675) was already working with these series before Maclaurin and Taylor were born. Also, the German born mathematician Nicolaus Mercator discovered the Maclaurin series for ln(1 + x) at just about the same time as Gregory. It seems, however, that Taylor was unaware of Gregory's work when he published his own book containing this series in 1715, and which Maclaurin quoted in his own work in 1742. Maclaurin never claimed to have first discovered the series named after him, but he actually was the originator of Cramer's Rule, which you may know as an algorithm used for solving systems of equations. In the end, it seems, we cannot always trust that the people whom theorems are named after are, indeed, the ones who should be celebrated for them!

Example 8

a Find the Maclaurin series generated by $f(x) = e^x$, and find its radius of convergence.
b Find the Maclaurin series generated by $f(x) = \cos(x)$.

a $f(x) = e^x$; $f^{(n)}(x) = e^x$; $f^{(n)}(0) = e^0 = 1$ for all n. $$e^x = 1 + x + \frac{x^2}{2!} + \ldots + \frac{x^n}{n!} + \ldots = \sum_{k=0}^{\infty} \frac{x^k}{k!}$$ $$\lim_{n \to \infty} \left| \frac{u_{n+1}}{u_n} \right| = \lim_{n \to \infty} \left| \frac{\frac{x^{n+1}}{(n+1)!}}{\frac{x^n}{n!}} \right| = \lim_{n \to \infty} \left| \frac{x^{n+1}}{(n+1)!} \cdot \frac{n!}{x^n} \right| = |x| \lim_{n \to \infty} \frac{1}{n+1} = 0$$ Since $L = 0 < 1$, the series converges, and $R = \frac{1}{L}$, or infinity. Hence, the series converges for all real x.	*Find successive derivatives of $f(x)$, and evaluate them at $x = 0$.* *Expand and write in general form.* *Use Ratio test for convergence.*
b $f(x) = \cos(x)$, $f(0) = 1$ $f'(x) = -\sin(x)$, $f'(0) = 0$ $f''(x) = -\cos(x)$, $f''(0) = -1$ $f'''(x) = \sin(x)$, $f'''(0) = 0$ $f^{(4)}(x) = \cos(x)$, $f^{(4)}(0) = 1$ $f^{(5)}(x) = -\sin(x)$, $f^{(5)}(0) = 0$ $$\cos x = 1 - \frac{x^2}{2!} + \frac{x^4}{4!} - \frac{x^6}{6!} + \ldots + (-1)^n \frac{x^{2n}}{(2n)!} + \ldots = \sum_{n=0}^{\infty} (-1)^n \frac{x^{2n}}{(2n)!}$$	*Find successive derivatives and evaluate them at $x = 0$. Observe that the derivatives repeat in cycles of 4, i.e. 1, 0, −1, 0, 1, 0, −1, 0, …* *Write out some of the terms of the Maclaurin series, and the general term.*

It should be noted here that if f can be represented by a power series centered at a, then f is equal to the sum of its Taylor series. There are, however, some functions that are not equal to the sum of their Taylor series. For example, it can be shown that $f(x) = \begin{cases} e^{-\frac{1}{x^2}}, & x \neq 0 \\ 0, & x = 0 \end{cases}$ is not equal to its

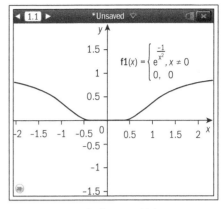

Taylor series.

Using the GDC to graph this function, we obtain

Using L'Hopital's Rule for series (which comes later in this chapter), it is possible to establish that every derivative of this function at $x = 0$ is simply 0, hence its Taylor series is simply 0, which clearly does not converge to this function. Therefore, infinite differentiability at a point is a necessary condition for a function to equal its Taylor series, but not a sufficient condition.

Under what conditions, therefore, is a function equal to the sum of its Taylor series? In other words, if f has derivatives of all orders, then when does $f(x) = \sum_{n=0}^{\infty} \frac{f^{(n)}(a)}{n!}(x-a)^n$? As with any convergent series, this would mean that $f(x)$ is the limit of the sequence of partial sums of its Taylor polynomials, i.e. $f(x) = \lim_{n \to \infty} T_n(x)$. If we therefore let $R_n(x) = f(x) - T_n(x)$, or $f(x) = T_n(x) + R_n(x)$, then $R_n(x)$ is called the remainder of the Taylor polynomial of order n. If $\lim_{n \to \infty} R_n(x) = 0$ (null polynomial), then $\lim_{n \to \infty} T_n(x) = \lim_{n \to \infty} (f(x) - R_n(x)) = f(x) - \lim_{n \to \infty} R_n(x) = f(x)$

We can now state the following theorem.

Theorem 5: If $f(x) = T_n(x) + R_n(x)$ where T_n is the nth order Taylor polynomial of f at $x = a$ and $R_n(x)$ is called the remainder of the Taylor polynomial of order n, and $\lim_{n \to \infty} R_n(x) = 0$ for $|x - a| < R$, then f is equal to the sum of its Taylor series on the interval $|x - a| < R$.

In attempting to show that $\lim_{n \to \infty} R_n(x) = 0$ for a specific function f, we use the following theorem:

Taylor's Formula: If f has $n + 1$ derivatives in the interval I that contains a, then there exists a real number t, $x < t < a$, such that the remainder term in the Taylor series can be expressed as $R_n(x) = \frac{f^{(n+1)}(t)}{(n+1)!}(x-a)^{n+1}$. This expression is referred to as **LaGrange's form** of the remainder term.

Note that in using the LaGrange form of the remainder term, we often estimate the error by replacing $|f^{(n+1)}(t)|$ with its maximum for $a \leq t \leq x$ when $x > a$, and $x \leq t \leq a$ when $a > x$.

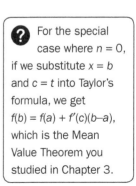

For the special case where $n = 0$, if we substitute $x = b$ and $c = t$ into Taylor's formula, we get $f(b) = f(a) + f'(c)(b-a)$, which is the Mean Value Theorem you studied in Chapter 3.

Chapter 5 149

We can now prove that some functions are equal to the sum of their Taylor series. For example, we will prove that $f(x) = e^x$ is equal to its Taylor series

$$e^x = 1 + x + \frac{x^2}{2!} + \ldots + \frac{x^n}{n!} + \ldots = \sum_{k=0}^{\infty} \frac{x^k}{k!}.$$

We first make use of the fact that $\lim_{n \to \infty} \frac{x^n}{n!} = 0$ for all real x, since we proved in Exercise 5B 1b that the series converges for all x, hence its nth term approaches 0.

Furthermore, if $f(x) = e^x$ then $f^{(n+1)}(x) = e^x$, and $R_n(x) = \frac{e^t}{(n+1)!} x^{n+1}$,

for $x > 0$, $0 < t < x$. Hence, $0 < \frac{x^{n+1}}{(n+1)!} e^t < \frac{x^{n+1}}{(n+1)!} e^x$. By the squeeze theorem, $\lim_{n \to \infty} R_n(x) = 0$

since $\lim_{n \to \infty} \frac{x^{n+1}}{(n+1)!} e^x = 0$ for all real x. If $x < 0$, then $x < t < 0$, and $e^t < e^0 = 1$, and again the result holds.

We now know that $e^x = \sum_{n=0}^{\infty} \frac{x^n}{n!}$ for all x. In particular it is interesting to note that when $x = 1$, $e^1 = e$, and we obtain another definition for e,

$$e = \sum_{n=0}^{\infty} \frac{1}{n!} = 1 + \frac{1}{1!} + \frac{1}{2!} + \frac{1}{3!} + \ldots + \frac{1}{n!} + \ldots$$

Example 9

Find the Maclaurin series for $f(x) = \sin(x)$, and prove that it represents the function for all x.

$f(x) = \sin(x)$, $\quad f(0) = 0$	Find successive derivatives of $f(x)$, and evaluate them at $x = 0$.
$f'(x) = \cos(x)$, $\quad f'(0) = 1$	
$f''(x) = -\sin(x)$, $\quad f''(0) = 0$	
$f'''(x) = -\cos(x)$, $\quad f'''(0) = -1$	
$f^{(4)}(x) = \sin(x)$, $\quad f^{(4)}(0) = 0$	
$\frac{1}{1!}x - \frac{1}{3!}x^3 + \frac{1}{5!}x^5 - \frac{1}{7!}x^7 + \ldots = \sum_{n=0}^{\infty} (-1)^n \frac{x^{2n+1}}{(2n+1)!}$	Write out the terms of the Maclaurin series, and its general term.
$R_n(x) = \frac{f^{(n+1)}(t)}{(n+1)!} x^{n+1}$, $f(x) = \sin(x)$ and $0 < t < x$. Since $f^{(n+1)}(t) = \pm\sin(x)$ or $\pm\cos(x)$, $|f^{(n+1)}(t)| \leq 1$.	Express R_n in terms of f, and evaluate $f^{(n+1)}(t)$.
$0 < \frac{\left|f^{(n+1)}(t)\right|}{(n+1)!} \left|x^{n+1}\right| \leq \frac{\left|x\right|^{n+1}}{(n+1)!}$	Show that $\lim_{n \to \infty} R_n(x) = 0$.
Since $\lim_{n \to \infty} \frac{\left|x\right|^{n+1}}{(n+1)!} = 0$, the squeeze theorem shows that $\lim_{n \to \infty} R_n(x) = 0$.	
Hence $\sin(x) = \sum_{n=0}^{\infty} (-1)^n \frac{x^{2n+1}}{(2n+1)!}$ for all x.	

Note that the properties on page 141 for power series also hold true for Taylor series (earlier, we stressed that this was the case for finite Taylor polynomials of order n, but here we remark that these properties hold for all infinite Taylor series). The Taylor series for the sum of two functions f and g is the sum of the Taylor series for f and g, and the product of the Taylor series for f and g is the Taylor series of the product of f and g. Hence, for example, if we want to find the Maclaurin series for $f(x) = \sin(x) + \cos(x)$, we simply find the Maclaurin series for $\sin(x)$ and for $\cos(x)$, and add them. The same holds for finding the Maclaurin series of $f(x) = \sin(x)\cos(x)$: We find the Maclaurin series for each function separately, and then find the product of the two series.

In addition, we can use substitution to find the Maclaurin series of $f(x) = \cos(2x)$. Since we know the Maclaurin expansion for $f(x) = \cos(x)$ is $\sum_{n=0}^{\infty}(-1)^n \frac{x^{2n}}{(2n)!}$, then substituting $2x$ for x, we obtain $\cos(2x) = \sum_{n=0}^{\infty}(-1)^n \frac{(2x)^{2n}}{(2n)!} = \sum_{n=0}^{\infty}(-1)^n \frac{(4x^2)^n}{(2n)!}$.

Using the properties of power series, we can find, for example, the Maclaurin series for $f(x) = \frac{2 + \cos 2x}{3}$. We simply use the Maclaurin expansion for $\cos(2x)$ from above, add 2 and divide 3.

Hence we obtain $f(x) = \frac{1}{3}\left(2 + \sum_{n=0}^{\infty}(-1)^n \frac{(4x^2)^n}{(2n)!}\right)$.

Example 10

Find the Maclaurin series for $f(x) = x \cos 2x$.

$x \cos 2x = x \sum_{n=0}^{\infty}(-1)^n \frac{(2x)^{2n}}{(2n)!}$	Substitute $2x$ for x in the Maclaurin expansion of $\cos(x)$, and multiply the whole expression by x.
$= \sum_{n=0}^{\infty}(-1)^n \frac{4^n x^{2n+1}}{(2n)!}$	Bring the x into the main expression in order to simplify.

Exercise 5E

1 Prove that the Maclaurin series for $\cos(x)$ represents $f(x) = \cos(x)$ for all x.

2 Find the Maclaurin expansion for
  a $\sin^2 x$
  b e^{3x}
  c $\ln(1-x)$
  d $\tan(x)$
  e $\arctan(x^2)$
  f $e^x \sin(x)$

3 Find the Maclaurin series for $\sin^2 x$ by using the identity $\sin^2 x = \frac{1}{2} - \frac{\cos 2x}{2}$. Comment on your result compared to the Maclaurin expansion you found for $\sin^2 x$ in **2a**.

4 Find the Maclaurin series for $f(x) = \frac{\sin x}{x}$, and use it to prove that $\lim_{x \to 0} \frac{\sin x}{x} = 1$.

5 Using the Maclaurin series for $\sin(x)$, $\cos(x)$, and e^x, prove

6 a By substituting $i\theta$ into the Taylor series for e^x, prove Euler's formula
$e^{i\theta} = \cos\theta + i\sin\theta$.

b From this formula, derive Euler's identity, $e^{\pi i} + 1 = 0$, and explain why you think it is referred to as the most beautiful equation ever written.

You will have noticed that direct computation of Maclaurin series can be very tedious. A practical way to find a Maclaurin series therefore is to start with a basic list of Maclaurin series for elementary functions, and from this list develop Maclaurin series for other functions using the arithmetic and calculus operations.

Before developing this list, we will complete our list of Maclaurin series for basic functions with two more examples.

Example 11

Find the Maclaurin series for $f(x) = \ln(x+1)$, and determine its radius of convergence.

$f(x) = \ln(x+1)$, $f(0) = 0$	Find successive derivatives of $f(x)$, and evaluate them at $x = 0$.
$f'(x) = \dfrac{1}{1+x}$, $f'(0) = 1$	
$f''(x) = -(1+x)^{-2}$, $f''(0) = -1$	
$f'''(x) = 2(1+x)^{-3}$, $f'''(0) = 2$	
$f^{(4)}(x) = -6(1+x)^{-4}$, $f^{(4)}(0) = -6$	
$\ln(x+1) = 0 + \dfrac{1}{1!}x - \dfrac{1}{2!}x^2 + \dfrac{2}{3!}x^3 - \dfrac{6}{4!}x^4 + \ldots$ $= x - \dfrac{x^2}{2} + \dfrac{x^3}{3} - \dfrac{x^4}{4} + \ldots + (-1)^{n-1}\dfrac{x^n}{n} + \ldots = \sum_{n=1}^{\infty}(-1)^{n-1}\dfrac{x^n}{n}$	Write out the terms of the Maclaurin series, and the general term.
$\lim_{n\to\infty}\left|\dfrac{\frac{x^{n+1}}{n+1}}{\frac{x^n}{n}}\right| = \lim_{n\to\infty}\left|x \cdot \dfrac{n+1}{n}\right| = \|x\|\lim_{n\to\infty}\left|\dfrac{n+1}{n}\right| = \|x\|.$	Apply the ratio test.
By the ratio test, $\|x\| < 1$, hence, $I = \,]-1, 1]$ and $R = 1$	

We will now find the Maclaurin series for $f(x) = (1+x)^p$, where p is a real number, and determine its radius of convergence. This is referred to as the Binomial Series.

By successive differentiation we obtain

$f(x) = (1+x)^p$	$f(0) = 1$
$f'(x) = p(1+x)^{p-1}$	$f'(0) = p$
$f''(x) = p(p-1)(1+x)^{p-2}$	$f''(0) = p(p-1)$
$f'''(x) = p(p-1)(p-2)(1+x)^{p-3}$	$f'''(0) = p(p-1)(p-2)$
….	….
$f^n(x) = p(p-1)(p-2)\ldots (p-(n-1))(1+x)^{p-n}$	$f^{(n)}(0) = p(p-1)(p-2)\ldots (p-(n-1))$

Everything polynomic

This produces the series
$$1 + px + \frac{p(p-1)}{2!}x^2 + \frac{p(p-1)(p-2)}{3!}x^3 + \ldots \frac{p(p-1)\ldots(p-n+1)}{n!}x^n + \ldots$$

By the ratio test,
$$\lim_{n\to\infty}\left|\frac{a_{n+1}}{a_n}\right| = |x|\lim_{n\to\infty}\left|\frac{p(p-1)\ldots(p-n)}{(n+1)!} \cdot \frac{n!}{p(p-1)\ldots(p-n+1)}\right|$$
$$= |x|\lim_{n\to\infty}\left|\frac{1}{n+1}\cdot(p-n)\right| = |x|$$

We know this converges for $|x| < 1$, hence the radius of convergence is 1. This tells us that this series converges to some function for x in the interval $]-1, 1[$. Convergence at the end points depends on the value of p. It turns out that the series converges at 1 if $-1 < p \leq 0$, and at both endpoints if $p > 0$.

To show that the series actually converges to the function $y = (1 + x)^p$, we can use techniques for solving differential equations.

We know from our table above that
$$y' = p + p(p-1)x + \frac{p(p-1)(p-2)}{2!}x^2 + \ldots + \frac{p(p-1)(p-2)\ldots(p-n+1)x^n}{n!} + \ldots$$

Multiplying both sides by x, we obtain
$$xy' = px + p(p-1)x^2 + \frac{p(p-1)(p-2)}{2!}x^3 + \ldots + \frac{p(p-1)(p-2)\ldots(p-n+2)x^n}{(n-1)!} + \ldots$$

Now adding y' and xy',
$$y' + xy' = p + [p(p-1) + p]x + \left[\frac{p(p-1)(p-2)}{2!} + p(p-1)\right]x^2 + \ldots$$
$$= p + p^2 x + \frac{p^2(p-1)x^2}{2!} + \ldots$$
$$= p\left(1 + px + \frac{p(p-1)x^2}{2!} + \ldots\right)$$
$$= py.$$

This gives us the differential equation $y' + xy' = py$, which in standard linear form is $y' - \frac{p}{1+x}y = 0$.

The general solution to this equation is therefore
$$ye^{\int g(x)dx} = A \Rightarrow y = A(1+x)^p \quad [\text{where } g(x) = -\frac{p}{1+x}]$$

Using the initial condition that $y = 1$ when $x = 0$, we see that $A = 1$ and the particular solution is therefore $y = (1 + x)^p$. So, we have shown that the function $f(x) = (1 + x)^p$ is equal to its Maclaurin series.

The Binomial series is the generalization of the Binomial Theorem when p is not an integer. When p is a positive integer and $n \leq p$, the expression $\binom{p}{n}$ contains a factor $(p - p)$ so $\binom{p}{n} = 0$. This means that the Binomial series terminates, and reduces to the Binomial expansion that you are familiar with, i.e., $(1+x)^p = \sum_{n=0}^{p} \binom{p}{n} x^n$, $1 \leq n \leq p$, where $\binom{p}{n} = \frac{p!}{n!(p-n)!} = \frac{p(p-1)(p-2)\ldots(p-n+1)}{n!}$.

Example 12

a Find the Maclaurin series for $f(x) = \dfrac{1}{(1+x)^2}$ using the Binomial series.

b Find the Maclaurin Series for $f(x) = \dfrac{1}{\sqrt{2-x}}$, and its radius of convergence.

---

a $\binom{-2}{n} = \dfrac{(-2)(-3)(-4)\ldots(-2-n+1)}{n!}$

$= (-1)^n \dfrac{2 \cdot 3 \cdot 4 \cdot \ldots \cdot n(n+1)}{n!} = (-1)^n (n+1)$

Use the binomial series with $p = -2$

Factor out -1

Hence, for $|x| < 1$

$\dfrac{1}{(1+x)^2} = \sum_{n=0}^{\infty} \binom{-2}{n} x^n = \sum_{n=0}^{\infty} (-1)^n (n+1) x^n$

$= 1 - 2x + 3x^2 - 4x^3 + 5x^4 + \ldots + (-1)^n (n+1) x^n + \ldots$

Substitute the expression found above, and generalize.

b $\dfrac{1}{\sqrt{2-x}} = \dfrac{1}{\sqrt{2\left(1-\frac{x}{2}\right)}} = \dfrac{1}{\sqrt{2}\sqrt{1-\frac{x}{2}}} = \dfrac{1}{\sqrt{2}}\left(1 - \dfrac{x}{2}\right)^{-\frac{1}{2}}$

Rewrite $f(x)$ in a form which allows us to use the Binomial series.

$= \dfrac{1}{\sqrt{2}} \sum_{n=0}^{\infty} \binom{-\frac{1}{2}}{n} \left(-\dfrac{x}{2}\right)^n$

Use the Binomial series with $p = -\dfrac{1}{2}$, and x replaced by $-\dfrac{x}{2}$. Although p is not an integer in this case, the formula applies.

$= \dfrac{1}{\sqrt{2}}\left[1 + -\dfrac{1}{2}\cdot -\dfrac{x}{2} + \dfrac{-\frac{1}{2}\cdot -\frac{3}{2}}{2!}\left(-\dfrac{x}{2}\right)^2 + \ldots + \right.$

$\left. \dfrac{\left(-\frac{1}{2}\right)\left(-\frac{3}{2}\right)\left(-\frac{5}{2}\right)\ldots\left(-\frac{2n-1}{2}\right)}{n!}\left(-\dfrac{x}{2}\right)^n + \ldots \right]$

$= \dfrac{1}{\sqrt{2}}\left[1 + \dfrac{1}{4}x + \dfrac{1 \cdot 3}{2! 4^2}x^2 + \dfrac{1 \cdot 3 \cdot 5}{3! 4^3}x^3 + \ldots\right.$

$\left. + \dfrac{1 \cdot 3 \cdot 5 \cdot \ldots \cdot (2n-1)}{n! 4^n}x^n + \ldots\right]$

Find an expression for the general term.

Interval of convergence is $\left|-\dfrac{x}{2}\right| < 1$, or $|-x| < 2$ hence $R = 2$.

Find R.

The following is a list of Maclaurin series for special functions, and can be found in the official IB higher level formula booklet. The general term of the series is included in the list below as well as the interval of convergence, but these are not included in the formula booklet.

Maclaurin Series	Interval of Convergence
$e^x = 1 + x + \dfrac{x^2}{2!} + \ldots + \dfrac{x^n}{n!} + \ldots$	All real x.
$\ln(1+x) = x - \dfrac{x^2}{2} + \dfrac{x^3}{3} + \ldots + (-1)^{n-1}\dfrac{x^n}{n} + \ldots$	$]-1, 1]$
$\sin x = x - \dfrac{x^3}{3!} + \dfrac{x^5}{5!} + \ldots + (-1)^n \dfrac{x^{2n+1}}{(2n+1)!} + \ldots$	All real x.
$\cos x = 1 - \dfrac{x^2}{2!} + \dfrac{x^4}{4!} + \ldots + (-1)^n \dfrac{x^{2n}}{(2n)!} + \ldots$	All real x.
$\arctan x = x - \dfrac{x^3}{3} + \dfrac{x^5}{5} + \ldots + (-1)^n \dfrac{x^{2n+1}}{2n+1} + \ldots$	$[-1, 1]$
* $(1+x)^p = 1 + px + \dfrac{p(p-1)x^2}{2!} + \ldots + \dfrac{p(p-1)\ldots(p-n+1)x^n}{n!} + \ldots$	$]-1, 1[$. Convergence at the endpoints is dependent on p.

*This Binomial series is not in the formula booklet.

Exercise 5F

1 Find the Maclaurin series for the following and state the interval of convergence.

   a $f(x) = \sqrt{1-x}$ **b** $f(x) = \dfrac{1}{(1+x)^3}$

   c $f(x) = \dfrac{1}{(1-4x^2)^2}$ **d** $f(x) = \dfrac{1}{\sqrt[4]{1+2x^3}}$

2 Find the Maclaurin series of $y = \arcsin(x)$ by expanding its derivative in a power series and integrating term by term.

3 Using the Binomial series, approximate the following to 4 s.f.

   a $\sqrt{9.1}$ **b** $\dfrac{1}{1.03^4}$

5.6 Using Taylor Series to approximate functions

Taylor Series can be used to approximate functions to a desired degree of accuracy. The approximations can be used to either evaluate functions at specific values of x, or to differentiate or integrate the function.

The example below illustrates how to use the truncation error studied in chapter 4 to approximate $\cos(0.1)$ within a desired level of accuracy.

Example 13

Approximate $\cos(0.1)$ with an error less than 10^{-20}.

Since 0.1 is close to 0, we can use the Maclaurin expansion for $f(x) = \cos(x) = \sum_{n=0}^{\infty}(-1)^n \dfrac{x^{2n}}{(2n)!}$.	Select the appropriate series.
Substituting 0.1 for x, $$\cos(0.1) = \sum_{n=0}^{\infty}(-1)^n \dfrac{0.1^{2n}}{(2n)!}.$$	Make the substitution.
$u_{n+1} = \dfrac{0.1^{2(n+1)}}{(2n+2)!} < 10^{-20}$ Using the GDC, we see that $n = 5$ is enough.	For an alternating series, the absolute value of the truncation error is less than the next term in the series. Use GDC to find n.
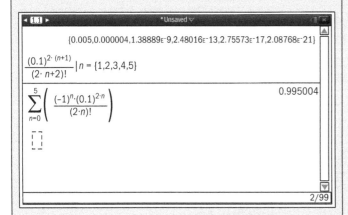	Use GDC to find the sum from $n = 0$ to $n = 5$.
Hence, $\cos(0.1)$ is approximately 0.99995.	

Similar to the truncation error for an alternating series, finding the error bound using Taylor's Formula for the remainder is given by the next term in the series, i.e. $R_n(x) = \dfrac{f^{(n+1)}(t)}{(n+1)!}(x-a)^{n+1}$.

The tricky part here though is that you evaluate the $(n + 1)$th derivative at t. We often estimate the error by replacing $\left|f^{(n+1)}(t)\right|$ with its maximum for $a \leq t \leq x$ when $x > a$, and $x \leq t \leq a$ when $a > x$. In other words, t is the number that makes $f^{n+1}(t)$ as large as can be. This error bound will tell you how far off you are from the real value, so you always want the error bound to represent the largest possible error. This may sound complicated, but in practice it's quite straightforward, as the following example will show.

Example 14

Approximate the value of ln(1.1) using a 3rd degree Taylor polynomial and determine the maximum error in this approximation.	
$T_3(x) = f(1) + f'(1)(x-1) + \dfrac{1}{2!}f''(1)(x-1)^2 + \dfrac{1}{3!}f'''(1)(x-1)^3$	1.1 is close to 1, so choose $x = 1$ as center
$f(x) = \ln(x)$ $f(1) = 0$ $f'(x) = \dfrac{1}{x}$ $f'(1) = 1$ $f''(x) = -\dfrac{1}{x^2}$ $f''(1) = -1$ $f'''(x) = \dfrac{2}{x^3}$ $f'''(1) = 2$	Evaluate successive derivatives of f at $x = 1$.
$T_3(x) = (x-1) - \dfrac{(x-1)^2}{2!} + \dfrac{2(x-1)^3}{3!}$	Substitute into $T_3(x)$ expressions and values found.
$T_3(1.1) = 0.1 - \dfrac{(0.1)^2}{2} + \dfrac{(0.1)^3}{3} \approx 0.0953333$	Evaluate $T_3(0.1)$
$R_3(x) = \dfrac{1}{4!}f^{(4)}(t)(x-1)^4 = \dfrac{1}{4!}\left(-\dfrac{6}{t^4}\right)(x-1)^4$ $a = 1; x = 1.1; 1 \leq t \leq 1.1$	Write out $R_3(x)$.
When $t = 1$, $R_3(1.1) = -\dfrac{1}{4(1)^4} \times 0.1^4 = -0.000025$	Determine which value of t will give the maximum error, i.e. the smaller t will give the larger fractional value.
Taking the absolute value, $R_3(1.1)$ is less than $0.000\,025$. Hence $\ln(1.1) \approx 0.0953333$ with an error of less than 0.000025.	$\ln(1.1) - 0.09533 = 0.00002$, which is less than 0.000025, as expected.

Let us consider again example 13, and apply the above method to find the maximum error in approximating $\cos(0.1)$. For the trigonometric functions $\cos(t)$ and $\sin(t)$, one need not consider the value of t that will maximize the error, rather the value of $\cos(t)$ and $\sin(t)$. Since the maximum value of both functions is 1, we use this to maximize the error, as shown in example below.

Example 15

Approximate $\cos(0.1)$ using a 4th degree Taylor polynomial and find the associated LaGrange remainder, or error bound.

$T_4(x) = 1 - \dfrac{x^2}{2!} + \dfrac{x^4}{4!}$	Write out $T_4(x)$, using $x = 0$ as the center.
$\cos(0.1) \approx 1 - \dfrac{0.1^2}{2!} + \dfrac{0.1^4}{4!} \approx 0.99500416667$	Substitute 0.1 for x into $T_4(x)$.
$R_4(x) = \dfrac{f^{(5)}(t)x^5}{5!} = \dfrac{(-\sin t)(0.1)^5}{5!}$ Let $-\sin(t) = 1$	Write out $R_4(x)$. Select the value that will maximize the error.
$R_4(0.1) = \dfrac{(1)(0.1)^5}{5!} \approx 8.33 \times 10^{-8}$	Substitute and evaluate.
$\cos(0.1) \approx 0.99500416667$ with an error less than 8.33×10^{-8}	$\cos(0.1) - 0.99500416667 \approx 1.39 \times 10^{-9}$ which is less than 8.33×10^{-8}.

If we are asked to compute $\int_0^1 e^{-x^2} dx$, we cannot do this using calculus methods, since e^{-x^2} does not have an anti-derivative that can be expressed as a simple function. We can, however, use its Maclaurin series representation, and obtain the expression below by substituting $-x^2$ for x in the Maclaurin expansion for e^x.

$$\int_0^1 e^{-x^2} dx = \int_0^1 \left(1 - x^2 + \frac{x^4}{2!} - \frac{x^6}{3!} + \ldots + (-1)^n \frac{x^{2n}}{n!} + \ldots \right) dx.$$

$$= \left[x - \frac{x^3}{3} + \frac{x^5}{5 \cdot 2!} - \frac{x^7}{7 \cdot 3!} + \ldots \right]_0^1$$

$$= 1 - \frac{1}{3} + \frac{1}{5 \cdot 2!} - \frac{1}{7 \cdot 3!} + \ldots$$

Using the GDC we can see that the more terms we add, the closer we get to the actual sum. Of course the GDC is also limited in the degree of accuracy that we can achieve.

$\int_0^1 e^{-x^2} dx$	0.746824
$1 - \frac{1}{3} + \frac{1}{10} - \frac{1}{7 \cdot 3!}$	0.742857
$1 - \frac{1}{3} + \frac{1}{10} - \frac{1}{7 \cdot 3!} + \frac{1}{9 \cdot 4!}$	0.747487
$1 - \frac{1}{3} + \frac{1}{10} - \frac{1}{7 \cdot 3!} + \frac{1}{9 \cdot 4!} - \frac{1}{11 \cdot 5!}$	0.746729
$\sum_{n=0}^{7} \left(\frac{(-1)^n}{(2 \cdot n + 1) \cdot n!} \right)$	0.746823
$\sum_{n=0}^{8} \left(\frac{(-1)^n}{(2 \cdot n + 1) \cdot n!} \right)$	0.746824

Let us now approximate $\int_0^1 e^{-x^2} dx$ with an error less than 0.001.

We need a series representation for this expression, and looking at the expanded series above, it is easy to see that

$$\int_0^1 e^{-x^2} dx = \sum_{x=0}^{\infty} (-1)^n \frac{1}{(2n+1)n!}.$$

This is an alternating series, and we know from chapter 4 that for an alternating series, the absolute value of the truncation error is less than the next term in the series.

Hence, $u_{n+1} = \frac{1}{(2n+3)(n+1)!} < 0.001$.

Using the GDC and trying several values for n, we see that the desired result is $n = 4$, and then evaluating the sum of the terms of the series from $n = 0$ to $n = 4$ gives us the approximation of $\int_0^1 e^{-x^2} dx$ with an error less than 0.001 to be 0.747, to 3 s.f.

$\frac{1}{(2 \cdot n + 3) \cdot (n+1)!} \mid n = \{1, 2, 3, 4\}$	{0.1, 0.02381, 0.00463, 0.000758}
$\sum_{n=0}^{4} \left((-1)^n \cdot \frac{1}{(2 \cdot n + 1) \cdot n!} \right)$	0.747487

Exercise 5G

1. Approximate e^{-1} within 0.001.

2. Use the power series expansion for $f(x) = \ln(x + 1)$ to estimate $\ln(1.5)$ with an error less than 0.0001.

3. Evaluate e to an error less than 0.00005 (or 4 decimal places).

4. Find the maximum error in approximating $\sin(x)$ by its Maclaurin series for $-\frac{1}{3} \leq x \leq \frac{1}{3}$.

5. Find the degree of the Taylor polynomial that should be used to approximate $e^{\frac{1}{3}}$ to four decimal places, and state the approximation.

6. Approximate $f(x) = x^{\frac{1}{3}}$ using a 2nd degree Taylor polynomial centered at $x = 8$, and determine the accuracy of this approximation when $7 \leq x \leq 9$.

7. Using a 3rd degree Taylor polynomial, evaluate $\int_0^{\frac{\pi}{4}} \sin^2 x \, dx$ to 6 d.p. Estimate the truncation error for the alternative series obtained, and compare this result with the GDC result for the definite integral.

8. Find the maximum error in using a 5th degree Taylor polynomial to approximate $\sin(0.5)$.

9. Find the maximum error in using a 4th degree Taylor polynomial to approximate $e^{0.2}$.

10. Find the degree of the Taylor polynomial necessary to approximate $\sin(1)$ within an error of 0.001.

5.7 Useful applications of power series

L'Hopital's rule and power series

At the beginning of the chapter, it states that some of the mathematics you have done up until now could be done more simply through the use of power series. You have seen several examples of this already, since it is sometimes easier to work with polynomials than it is to work with the given function, e.g. $y = e^{-x^2}$.

We could evaluate $\lim_{x \to 0} \dfrac{e^x - 1 - x}{x^2}$ using L'Hopital rule, since the limit of both the numerator and denominator is zero,

i.e. $\lim_{x \to 0} \dfrac{e^x - 1 - x}{x^2} = \dfrac{e^0 - 1 - 0}{0} = \dfrac{0}{0}$. Alternatively, we could use the Maclaurin series expansion for e^x, since power series are continuous functions, and obtain

$$\lim_{x \to 0} \dfrac{e^x - 1 - x}{x^2} = \lim_{x \to 0} \dfrac{\left(1 + \dfrac{x}{1!} + \dfrac{x^2}{2!} + \ldots\right) - 1 - x}{x^2}$$

$$= \lim_{x \to 0} \dfrac{\dfrac{x^2}{2!} + \dfrac{x^3}{3!} + \dfrac{x^4}{4!} + \ldots}{x^2}$$

$$= \lim_{x \to 0} \left(\dfrac{1}{2!} + \dfrac{x}{3!} + \dfrac{x^2}{4!} + \ldots\right) = \dfrac{1}{2}$$

Note that you only need to add as many terms as necessary in order to find the limit; that is, up to terms in x since these will vanish as x approaches 0.

In the end you should choose which method is more accessible. The rule of thumb, however, is that if it's going to take multiple tries of L'Hopital rule to find the limit, it might be best to use power series, if you can replace the function with a power series.

Example 16

Evaluate $\lim_{x \to 0} \dfrac{\cos x - 1 + \dfrac{x^2}{2}}{x^4}$.

$\lim_{x \to 0} \dfrac{\cos x - 1 + \dfrac{x^2}{2}}{x^4} = \lim_{x \to 0} \dfrac{\left(1 - \dfrac{x^2}{2!} + \dfrac{x^4}{4!} + \ldots + (-1)^n \dfrac{x^{2n}}{(2n)!} + \ldots\right) - 1 + \dfrac{x^2}{2}}{x^4}$	Substitute the Maclaurin expansion for $\cos(x)$.
$= \lim_{x \to 0} \dfrac{\dfrac{x^4}{4!} + \ldots + (-1)^n \dfrac{x^{2n}}{(2n)!} + \ldots}{x^4}$	Simplify and evaluate.
$= \lim_{x \to 0} \dfrac{1}{4!} - \dfrac{x^6}{6! x^4} + \ldots + (-1)^n \dfrac{x^{2n}}{(2n)! x^4} + \ldots$	
$= \dfrac{1}{4!}$ or $\dfrac{1}{24}$.	

Differential equations and Taylor series

Power series can be used to make the solution of certain types of differential equations more accessible. A power series represents a function f on an interval of convergence, and successively differentiating the power series generates series for f', f'', etc.

Let us consider approximating the solution of $y' = y^2 - x$ on the interval $[0, 1]$ using the first six terms of a Maclaurin series, given the initial condition that $y = 1$ when $x = 0$.

> In IB exams, the notation $\frac{dy}{dx}$ and $\frac{d^2y}{dx^2}$ will be used instead of y' and y'', respectively.

The general form of a Maclaurin series is

$$f(x) = f(0) + f'(0)x + \frac{f''(0)}{2!}x^2 + \frac{f'''(0)}{3!}x^3 + \ldots$$

Using the initial condition and differential equation, we obtain

	$y(0) = 1$
$y' = y^2 - x$	$y'(0) = 1$
$y'' = 2yy' - 1$	$y''(0) = 1$
$y''' = 2yy'' + 2(y')^2$	$y'''(0) = 4$
$y^{(4)} = 2yy''' + 6y'y''$	$y^{(4)}(0) = 14$
$y^{(5)} = 2yy^{(4)} + 8y'y''' + 6(y'')^2$	$y^{(5)}(0) = 66$

Substituting these into the general form of a Maclaurin series,

$$f(x) = 1 + x + \frac{1}{2}x^2 + \frac{4}{3!}x^3 + \frac{14}{4!}x^4 + \frac{66}{5!}x^5 + \ldots$$

We can now use this series to approximate values of y in the interval $[0, 1]$, as shown in the GDC screen shot.

x	f1(x):= 1+x+1/2*...
0.	1.
0.1	1.10573
0.2	1.22644
0.3	1.36906
0.4	1.54323
0.5	1.76198
0.6	2.04237
0.7	2.40616
0.8	2.88049
0.9	3.49849
1.	4.3
1.1	5.33217
1.2	6.65018
1.3	8.31784

Everything polynomic

We will now consider an exam-style question.

Example 17

> Given the differential equation $\frac{dy}{dx} - y\tan x = \cos x$ and the initial condition $\left(0, -\frac{\pi}{2}\right)$, find the Maclaurin series for y up to and including the term in x^2.

	$y(0) = -\frac{\pi}{2}$	Evaluate $y(0)$, $y'(0)$ and $y''(0)$.
$y' - y\tan x = \cos x$	$y'(0) = 1$	
$y'' = y\sec^2 x + y'\tan x - \sin x$	$y''(0) = -\frac{\pi}{2}$	
$y = -\frac{\pi}{2} + x - \frac{\pi}{4}x^2$		Substitute values into the form for a Maclaurin series.

Exercise 5H

1 Compute the following limits using Taylor Series

  a $\lim\limits_{x \to 0} \dfrac{\sin x - x}{x^3}$ **b** $\lim\limits_{x \to 0} \dfrac{e^x - e^{-x}}{x}$

  c $\lim\limits_{x \to 0} \dfrac{(2 - 2\cos x)^3}{x^6}$ **d** $\lim\limits_{x \to 0} \dfrac{x^2 - \sin^2 x}{x^2 \sin^2 x}$

  e $\lim\limits_{x \to 0} \dfrac{(\sin 3x - 3x)^2}{(\cos 5x - 1)^3}$

2 Find a Taylor series to approximate the solution of $\dfrac{dy}{dx} = y^2 - x$ given the initial condition that $y = 1$ when $x = 0$. Using the first six terms of the series, approximate the value of y when $x = 0.2$

3 Approximate the solution of the differential equation $\dfrac{dy}{dx} = 3x^2 y$ using a power series with four terms, such that $y = 1$ at $x = 1$.

4 Obtain a suitable power series to approximate the solution of the differential equation $x^2 \dfrac{dy}{dx} = y - x - 1$, and explain why the power series is not a solution for the given differential equation.

Review exercise

EXAM-STYLE QUESTIONS

1. Find the radius of convergence of the series $\sum_{n=1}^{\infty} \dfrac{(2n-2)!}{n!(n-1)!} x^n$.

2. Let $f(x) = e^{-x} \cos 2x$.
   a. Show that:
      i. $f''(x) + 2f'(x) + 5f(x) = 0$
      ii. $f^{(n+2)}(0) + 2f^{(n+1)}(0) + 5f^{(n)}(0) = 0$.
   b. Find the Maclaurin series for $f(x)$ up to an including the term in x^4.

3. a. By stating the necessary conditions, show that the integral test can be used to determine the convergence of the series $\sum_{n=1}^{\infty} \dfrac{n}{e^{n^2}}$, and determine whether or not the series converges.
   b. Find the first four terms of the Maclaurin series for $\sin(x)$ and e^{x^2}.
   c. Find the Maclaurin series for $f(x) = \sin(x)\, e^{x^2}$ up to the term in x^5.
   d. Use part (c) to find $\lim\limits_{x \to 0} \dfrac{e^{x^2} \sin x - x}{x^3}$.

4. Let $f(x) = \arcsin(x)$ for $|x| \leq 1$. The derivatives of f satisfy the equation
   $(1-x^2) f^{(n+2)}(x) - 2(n+1)x f^{(n+1)}(x) - n^2 f^{(n)}(x) = 0$, $n \geq 1$.
   Assuming that the Maclaurin series for $f(x)$ contains only odd powers of x,
   a. i. show that for $n \geq 1$, $(n+1)(n+2) a_{n+2} = n^2 a_n$.
      ii. find an expression for a_n which is valid for odd $n \geq 3$, if $a_1 = 1$.
   b. Find the radius of convergence of the Maclaurin series.
   c. If $x = \dfrac{1}{2}$, find an approximate value for π to 4 s.f.

5. Find the interval of convergence, including endpoints, for the series $\sum_{n=1}^{\infty} \dfrac{(x-3)^n}{2^n \sqrt{n}}$.

6. Use the Maclaurin series for e^x to estimate $e^{0.2}$ correct to 3 d.p. with an error term less than 0.0005.

7. Find the first three non-zero terms of the expansion $g(x) = \dfrac{1}{2} \ln\left(\dfrac{1+x}{1-x}\right)$.

8. Find the Maclaurin series of $f(x) = \sqrt[3]{1+x}$ up to the term in x^3, and use it to approximate $\sqrt[3]{1.2}$ to 5 d.p.

9 Use the Maclaurin expansion of $\sin(x)$ to estimate $\sin(3°)$ to 5 d.p.

10 Let $f(x) = \dfrac{e^x + e^{-x}}{2}$.

   a Obtain an expression for $f^{(n)}(x)$.
   b Hence, derive the Maclaurin series for $f(x)$ up to the term in x^4.
   c Using your result from part (b), find a rational approximation to $f\left(\dfrac{1}{2}\right)$, and use the LaGrange error term to find an upper bound for the error in this approximation.

11 Let $f(x) = \ln(1 + \sin(x))$

   a Show that $f''(x) = -\dfrac{1}{1+\sin x}$.
   b Find the Maclaurin series for $f(x)$ up to the term in x^4.
   c Determine the Maclaurin series for $\ln(1 - \sin(x))$ up to x^4.
   d Using the series found above, show that $\ln(\sec x) = \dfrac{x^2}{2} + \dfrac{x^4}{12} + \ldots$
   e Hence, or otherwise, find $\displaystyle\lim_{x \to 0} \dfrac{\ln \sec x}{x\sqrt{x}}$.

12 Use power series to show that the solution of the initial value problem $y' = y^2 + 1$, $y(0) = 0$, is $y(x) = \tan(x)$.

13 Given that $\dfrac{dy}{dx} - y\tan x = \cos x$ and $y = -\dfrac{\pi}{2}$ when $x = 0$,

   a Find the Maclaurin series for y up to the term in x^2.
   b Solve the differential equation given that $y = 0$ when $x = \pi$. Give the solution in the form $y = f(x)$.

14 Find the Binomial series expansion for $f(x) = \sqrt{\dfrac{1+x}{1-2x}}$ up to and including the term in x^3, and state the interval of convergence.

15 Given that $f(x) = e^{(e^x - 1)}$, find the Maclaurin series for $f(x)$, and hence find the value of $\displaystyle\lim_{x \to 0} \dfrac{f(x) - 1}{f'(x) - 1}$.

16 Determine the number of terms needed in the series $\dfrac{\pi}{4} = 1 - \dfrac{1}{3} + \dfrac{1}{5} - \dfrac{1}{7} + \ldots$ to estimate π to four decimal places.

17 Approximate $P(0 < x < 1)$ with an error less than 0.0001 for the standard normal probability density function $P(a < x < b) = \dfrac{1}{\sqrt{2\pi}} \displaystyle\int_a^b e^{-\frac{x^2}{2}} dx$.

Chapter 5 summary

Definitions

1. An expression of the form $\sum_{n=0}^{\infty} c_n x^n = c_0 + c_1 x + c_2 x^2 + \ldots + c_n x^n + \ldots$, c a real number, is a **power series centered at $x = 0$**.

   The set of points for which the series converges is the **interval of convergence**, and this set consists of the interval $]-R, R[$ where R is the **radius of convergence**.

2. An expression of the form
   $$\sum_{n=0}^{\infty} c_n (x-a)^n = c_0 + c_1 (x-a) + c_2 (x-a)^2 + \ldots + c_n (x-a)^n + \ldots,$$ a, c and a real numbers, is a power series centered at $x = a$. The set of points for which the series converges is the **interval of convergence**, and this set consists of the interval $]a - R, a + R[$ where R is the **radius of convergence**.

3. If $f(x)$ has n derivatives at $x = a$, then the taylor polynomial of degree n for $f(x)$ centered at a, $T_n(x)$, is the unique polynomial of degree n which satisfies
   $T_n(a) = f(a); T_n'(a) = f'(a); T_n''(a) = f''(a); \ldots; T^{(n)}(a) = f^{(n)}(a)$.

   Furthermore,
   $$T_n(x) = f(a) + \frac{f'(x)}{1!}(x-a) + \frac{f''(x)}{2!}(x-a)^2 + \ldots + \frac{f^{(n)}(a)}{n!}(x-a)^n = \sum_{k=0}^{n} \frac{f^{(k)}(a)}{k!}(x-a)^k.$$

4. $f(x)$ has derivatives of all orders throughout an open interval I such that $a \in I$, then the Taylor Series generated by f at $x = a$ is
   $$f(a) + f'(a)(x-a) + \frac{f''(a)}{2!}(x-a)^2 + \frac{f'''(a)}{3!}(x-a)^3 + \ldots = \sum_{k=0}^{\infty} \frac{f^{(k)}(a)}{k!}(x-a)^k.$$

   If $a = 0$, then the Taylor Series is also referred to as the **Maclaurin series** generated by f. The partial sums of both the Taylor and Maclaurin Series have already been defined, and are $T_n(x) = \sum_{k=0}^{n} \frac{f^{(k)}(a)}{k!}$.

Theorems

1. For a power series $f(x) = \sum_{n=0}^{\infty} c_n(x-a)$ centered at a, either
   - The series converges only at a.
   - The series converges for all x.
   - The series converges for $|x - a| < R$, where $R > 0$.

2. **Ratio Test Applied to Power Series**: For a power series
   $f(x) = \sum_{n=0}^{\infty} c_n (x-a)^n$ if
   - $\lim_{n \to \infty} \left| \frac{a_{n+1}}{a_n} \right| = L, L \neq 0 \Rightarrow R = \frac{1}{L}$
   - $\lim_{n \to \infty} \left| \frac{a_{n+1}}{a_n} \right| = 0$, then the radius of convergence is infinite
   - $\lim_{n \to \infty} \left| \frac{a_{n+1}}{a_n} \right| = \infty \Rightarrow R = 0$.

3 If $f(x) = \sum_{n=0}^{\infty} c_n(x-a)^n = c_0 + c_1(x-a) + c_2(x-a)^2 + \ldots + c_n(x-a)^n + \ldots$

converges in the interval $|x-a| < R$, R a real number, then the series

$f'(x) = \sum_{n=1}^{\infty} nc_n(x-a)^{n-1}$ obtained by differentiating the original series term by term

also converges in the interval $|x-a| < R$. Furthermore, if the series for f converges for all x, then so does the series for f'.

4 If $f(x) = \sum_{n=0}^{\infty} c_n(x-a)^n$ converges for $|x-a| < R$, R a real number, then the series

$\sum_{n=0}^{\infty} c_n \frac{(x-a)^{n+1}}{n+1}$ obtained by integrating f term by term also converges on the same

interval and represents $\int_a^x f(t)\,dt$ on that interval. Furthermore, if the series for f

converges for all x, then the series for the integral does as well.

5 Given the power series represented by the functions f and g, i.e.,

$f(x) = c_n(x-a)^n$ and $g(x) = d_n(x-a)^n$, the following properties hold:

- $f(kx) = \sum_{n=0}^{\infty} c_n k^n (x-a)^n$

- $f(x) \pm g(x) = \sum_{n=0}^{\infty} (c_n \pm d_n)(x-a)^n$

- $f(x) \cdot g(x) = \left(\sum_{n=0}^{\infty} c_n(x-a)^n\right)\left(\sum_{n=0}^{\infty} d_n(x-a)^n\right)$

Furthermore, let the radius of convergence of f be R_f and g be R_g. Then
- Radius of convergence of the sum when $R_f \neq R_g$ is $\min\{R_f, R_g\}$, i.e., the minimum of the two radii, or greater than or equal to the radii if $R_f = R_g$.
- Radius of convergence of the product is $\min\{R_f, R_g\}$, unless either $f(x) = 0$ or $g(x) = 0$.

6 If $f(x) = T_n(x) + R_n(x)$ where T_n is the nth order Taylor polynomial of f at $x = a$, $\lim_{n \to \infty} R_n(x) = 0$ for $|x-a| < R$, then f is equal to the sum of its Taylor series on the interval $|x-a| < R$.

In attempting to show that $\lim_{n \to \infty} R_n(x) = 0$ for a specific function f, we use the following Theorem.

7 Taylor's Formula: If f has $n+1$ derivatives in the interval I that contains a, there exists a real number t, t is between x and a, such that the remainder term in the Taylor

series can be expressed as $R_n(x) = \frac{f^{(n+1)}(t)}{(n+1)!}(x-a)^{n+1}$. This expression is referred to as

LaGrange's form of the remainder term.

Note that in using the LaGrange form for the remainder term, we often estimate the error by replacing $\left|f^{(n+1)}(t)\right|$ with its maximum for $a \leq t \leq x$ when $x > a$, and $x \leq t \leq a$ when $a > x$.

Answers

Chapter 1

Skills check

1 a $\dfrac{-6}{16n^2-1}$ **b** $-\dfrac{3}{n^2+5n+6}$

2 a $0 \le x \le \dfrac{2}{3}$ **b** $x \le -\dfrac{1}{2}$ or $x \ge \dfrac{7}{2}$

3 a $\dfrac{2n(\sqrt{n}-1)}{n-1}$ **b** $(n-1)(\sqrt{n+1}-\sqrt{n})$

4 a $\dfrac{(-1)^{n+1}}{2^n}$ **b** $\dfrac{(-1)^{n+1}}{n+1}$

c $\dfrac{2n-1}{2n}$ **d** $(-1)^{n+1}\sqrt{\dfrac{2n+1}{2^n}}, n \in \mathbb{Z}^+$

Investigation 1

1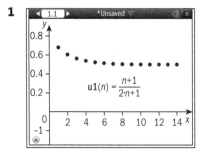

2 As n increases the terms get very close to $\dfrac{1}{2}$.

3 $m = 3$

4 $m = 25$, $m = 250$ and $m = 2500$.

5 No, as this implies that $n < -\dfrac{1}{8}$.

6 As n increases the terms get very close to 0.

7 $m = 5$, $m = 7$, and $m = 9$.

8 $|u_n - L|$ takes arbitrary small values as we consider only the terms when $n > m$.

Exercise 1A

1 $m = 1250$

2 a 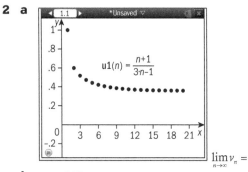 $\lim\limits_{n \to \infty} v_n = \dfrac{1}{3}$

b $m = 445$

3 a 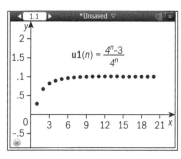 $\lim\limits_{n \to \infty} u_n = 1$

b $m = 7$

4 a Diverges.
b Converges to zero.
c Diverges.
d Converges to zero.
e Diverges to positive infinity.
f Diverges.
g Diverges.
h Converges to $\dfrac{1}{2}$.

Exercise 1B

1 a Proof **b** Proof

2 a Proof **b** $\dfrac{3}{4}$ **c** $\dfrac{3}{4}$

3 a Proof **b** 0

4 Assuming $\lim\limits_{n \to \infty} a_n$ exists, then $\lim\limits_{n \to \infty} a_n = 1$

Investigation 2

1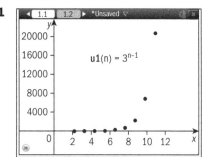

2 The terms of the sequence increase without upper bound.

3 $m = 6$

4 $m = 8$, $m = 10$ and $m = 12$

5 $m = 10$, $m = 14$ and $m = 17$

6 The terms of the sequence decrease without lower bound.

7 $m = 5$, $m = 7$ and $m = 9$

8 The absolute value of the terms of the sequence increase without upper bound.

Investigation 3

a 0 **b** 0 **c** 1 **d** 1 **e** $+\infty$

Exercise 1C

1. **a** 2.5 **b** 0 **c** $+\infty$ **d** $+\infty$
   e 0 **f** 1 **g** 0 **h** $+\infty$
   i 1 **j** $-\infty$ **k** $\dfrac{1}{2}$ **l** 0

2. **a** Proof **b** 0

Exercise 1D

1. **b**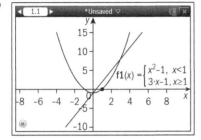

2. **a**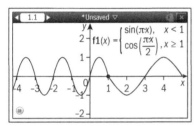

   $\lim\limits_{x\to 1^-} f(x) = \lim\limits_{x\to 1^+} f(x) = 0$ so $\lim\limits_{x\to 1} f(x) = f(1) = 0$

   b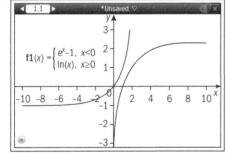

   $\lim\limits_{x\to 0^-} g(x) = 0$, $\lim\limits_{x\to 0^+} g(x) = \infty$ so $\lim\limits_{x\to 0} g(x)$ does not exist

   c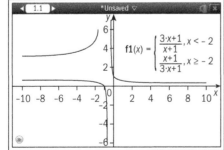

   $\lim\limits_{x\to -2^-} f(x) = 5$, $\lim\limits_{x\to -2^+} f(x) = \dfrac{1}{5}$, so $\lim\limits_{x\to -2} f(x)$ does not exist.

3. **a** 0 **b** 5 **c** ∞

4. **a**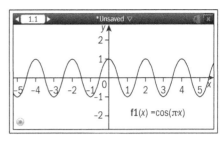

Review exercise

1. **a** $\dfrac{1}{2}$ **b** 0 **c** 0
   d 0 **e** 0 **f** 0

2. e.g. $u_n = -\dfrac{1}{n}$

3. 0

4. Proof

5. **a** $\dfrac{1}{3n(3n-1)}, n \in \mathbb{Z}^+$ **b** 1 **c** 0

6. 0

7. **c** $a_n = \left(\dfrac{1}{2}\right)^{n-2} + 1 \to 1$ **d** 1

Chapter 2
Before you start
You should know how to

1.

2.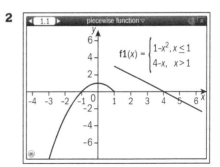

Skills check

1 a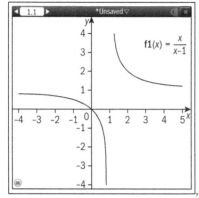

asymptotes: $x = 1$, $y = 1$

b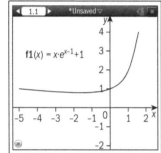

asymptote: $y = 1$

c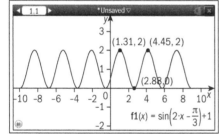

no asymptotes

d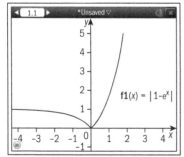

asymptote: $y = 1$; minimum at $x = 0$; axes intercept at origin

e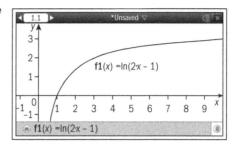

asymptote: $x = \dfrac{1}{2}$; no max/min; x-intercept: 1

2 a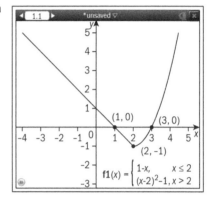

no asymptotes

b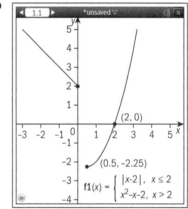

no asymptotes

c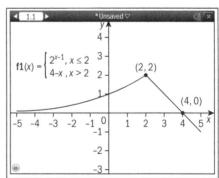

asymptote: $y = 0$

3 a 4   **b** -8

Exercise 2A

1 a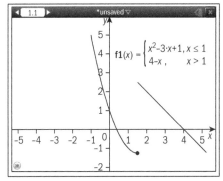

discontinuous at $x = 1$

b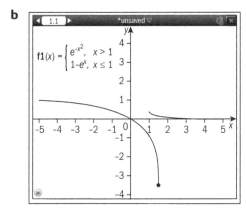

discontinuous at $x = 1$

c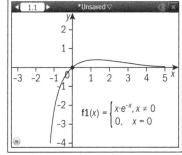

continuous in $\mathbb{R}$

d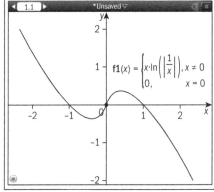

continuous in $\mathbb{R}$

Investigation - Composition of continuous functions

c g is continuous at $x = a$ and f is defined and continuous at $x = g(a)$.

Exercise 2B

a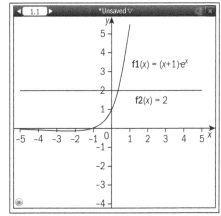

Apply Theorem 5 (Bolzano's) to a closed interval that contains the point of intersection of the graphs (e.g. [0.1]).

b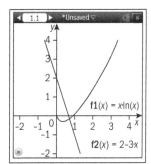

Apply Theorem 5 (Bolzano's) to a closed interval that contains the point of intersection of the graphs (e.g. [0, 1]).

c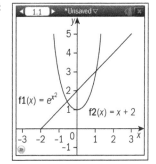

Apply theorem 5 (Bolzano's) to closed intervals that contain the points of intersection of the graphs (e.g. [–1, 0] and [0, 2]).

Answers 171

2 $f(x) = x$

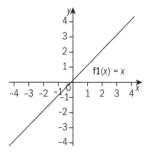

$f(x) = -x$

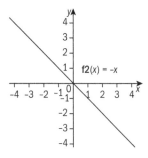

$f(x) = |x|$

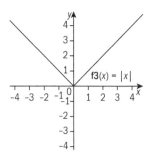

$f(x) = -|x|$

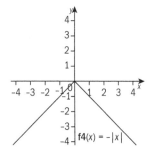

Reasons: $(f(x))^2 = x^2 \Rightarrow f(x) = \pm x$ and, as f is continuous, the function can just change from $f(x) = x$ to $f(x) = -x$ at $x = 0$, so we can also have $f(x) = |x|$ and $f(x) = -|x|$.

3 c can be any rational number.

Exercise 2C

1 b $f'(x) = \begin{cases} 4, & x < 0 \\ -2, & x < 0 \end{cases}$

c

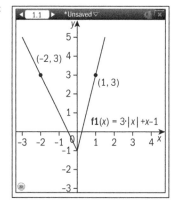

This function is not differentiable at $x = 0$, so we cannot use Rolle's Theorem.

2 a $f(-1) = 13$ and $f(1) = -3$

b $f'(x) = 10x^4 - 20x^3 - 30x^2$; $-1, 0$ and 3

c

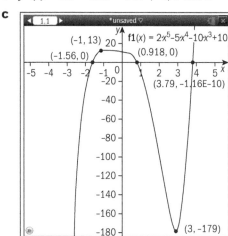

3 c

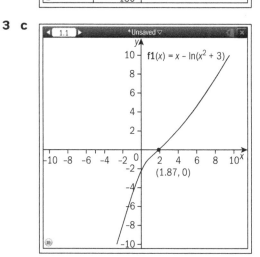

Investigation

a If using Microsoft Excel, create a spreadsheet similar to this format (with formulas shown below) to model the accuracy of the approximation $f(x_2) \approx f(x_1) + (x_2 - x_1)f'(x_1)$. Here we just show the case where $h = 1$.

				$f(x) = \sqrt{x}$, $h = 1$			
x_1	$f(x_1)$	$f'(x_1)$	h	x_2	$f(x_1) + (x_2 - x_1)f'(x_1)$	$f(x_2)$	% error
2	1.414	0.707	1	3	2.121	1.732	22.47%
3	1.732	0.577	1	4	2.309	2.000	15.47%
5	2.236	0.447	1	6	2.683	2.449	9.54%
7	2.646	0.378	1	8	3.024	2.828	6.90%
10	3.162	0.316	1	11	3.479	3.317	4.88%

				$f(x) = \sqrt{x}$, $h = 1$			
x_1	$f(x_1)$	$f'(x_1)$	h	x_2	$f(x_1) + (x_2 - x_1)f'(x_1)$	$f(x_2)$	% error
2	= SQRT(A4)	= 1/SQRT(A4)	1	= A4 + D4	= B4 + (E4 − A4)*C4	= SQRT(E4)	= ((F4 − G4)/G4)
3	= SQRT(A5)	= 1/SQRT(A5)	1	= A5 + D5	= B5 + (E5 − A5)*C5	= SQRT(E5)	= ((F5 − G5)/G5)
5	= SQRT(A6)	= 1/SQRT(A6)	1	= A6 + D6	= B6 + (E6 − A6)*C6	= SQRT(E6)	= ((F6 − G6)/G6)
7	= SQRT(A7)	= 1/SQRT(A7)	1	= A7 + D7	= B7 + (E7 − A7)*C7	= SQRT(E7)	= ((F7 − G7)/G7)
10	= SQRT(A8)	= 1/SQRT(A8)	1	= A8 + D8	= B8 + (E8 − A8)*C8	= SQRT(E8)	= ((F8 − G8)/G8)

The percentage error in the approximation decreases as h decreases, and as x_1 increases.

b If using Microsoft Excel, create a spreadsheet similar to this format (with formulas shown below) to model the accuracy of the approximation $f(x_2) \approx f(x_1) + (x_2 - x_1)f'(x_1)$. Here we just show the case where $h = 0.1$.

				$f(x) = e^x$, $h = 0.1$			
x_1	$f(x_1)$	$f'(x_1)$	h	x_2	$f(x_1) + (x_2 - x_1)f'(x_1)$	$f(x_2)$	% error
2	7.389	7.389	0.1	2.1	8.128	8.166	0.47%
3	20.086	20.086	0.1	3.1	22.094	22.198	0.47%
5	148.413	148.413	0.1	5.1	163.254	164.022	0.47%
7	1096.633	1096.633	0.1	7.1	1206.296	1211.967	0.47%
10	22026.466	22026.466	0.1	10.1	24229.112	24343.009	0.47%

$f(x) = e^x$, $h = 0.1$

x_1	$f(x_1)$	$f'(x_1)$	h	x_2	$f(x_1) + (x_2 - x_1)f'(x_1)$	$f(x_2)$	% error
2	= EXP(A13)	= EXP(A13)	0.1	= A13 + D13	= B13 + (E13 − A13)*C13	= EXP(E13)	= ((G13 − F13)/G13)
3	= EXP(A14)	= EXP(A14)	0.1	= A14 + D14	= B14 + (E14 − A14)*C14	= EXP(E14)	= ((G14 − F14)/G14)
5	= EXP(A15)	= EXP(A15)	0.1	= A15 + D15	= B15 + (E15 − A15)*C15	= EXP(E15)	= ((G15 − F15)/G15)
7	= EXP(A16)	= EXP(A16)	0.1	= A16 + D16	= B16 + (E16 − A16)*C16	= EXP(E16)	= ((G16 − F16)/G16)
10	= EXP(A17)	= EXP(A17)	0.1	= A17 + D17	= B17 + (E17 − A17)*C17	= EXP(E17)	= ((G17 − F17)/G17)

The percentage error in the approximation decreases as h decreases, but does not seem to depend on x_1.

c If using Microsoft Excel, create a spreadsheet similar to this format (with formulas shown below) to model the accuracy of the approximation $f(x_2) \approx f(x_1) + (x_2 - x_1) f'(x_1)$. Here we just show the case where $h = 0.01$.

$f(x) = \ln x$, $h = 0.01$

x_1	$f(x_1)$	$f'(x_1)$	h	x_2	$f(x_1) + (x_2 - x_1)f'(x_1)$	$f(x_2)$	% error
2	0.693	0.500	0.01	2.01	0.698	0.698	0.00178%
3	1.099	0.333	0.01	3.01	1.102	1.102	0.00050%
5	1.609	0.200	0.01	5.01	1.611	1.611	0.00012%
7	1.946	0.143	0.01	7.01	1.947	1.947	0.00005%
10	2.303	0.100	0.01	10.01	2.304	2.304	0.00002%

$f(x) = \ln x$, $h = 0.01$

x_1	$f(x_1)$	$f'(x_1)$	h	x_2	$f(x_1) + (x_2 - x_1)f'(x_1)$	$f(x_2)$	% error
2	= LN(A22)	= 1/A22	0.01	= A22 + D22	= B22 + (E22 − A22)*C22	= LN(E22)	= ((F22 − G22)/G22)
3	= LN(A23)	= 1/A23	0.01	= A23 + D23	= B23 + (E23 − A23)*C23	= LN(E23)	= ((F23 − G23)/G23)
5	= LN(A24)	= 1/A24	0.01	= A24 + D24	= B24 + (E24 − A24)*C24	= LN(E24)	= ((F24 − G24)/G24)
7	= LN(A25)	= 1/A25	0.01	= A25 + D25	= B25 + (E25 − A25)*C25	= LN(E25)	= ((F25 − G25)/G25)
10	= LN(A26)	= 1/A26	0.01	= A26 + D26	= B26 + (E26 − A26)*C26	= LN(E26)	= ((F26 − G26)/G26)

The percentage error in the approximation decreases as h decreases, and decreases as x_1 increases.

Exercise 2D

1. **a** -3 **b** 5 **c** $\frac{2}{3}$
   d $+\infty$ **e** 1 **f** 0
   g 0 **h** $+\infty$ **i** 0
3. **a** $+\infty$ **b** 0 **c** 1
4. **b** Using GDC, solve the polynomial $x^4 - 2x^3 + x^2 - 4 = 0$ to obtain solutions $x = -1$ and $x = 2$. Thus, one solution on $]1, 3[$.

Exercise 2E

1. **a** $\infty \times 0; 0$ **b** $\infty \times 0; \infty$ **c** $\infty \times 0; 1$
   d $1^\infty; e^4$ **e** $0^0; 1$ **f** $1^\infty; e^5$
   g $\infty \times 0; \frac{2}{5}$ **h** $1^\infty; e^{-\frac{1}{e}}$. **i** $1^\infty; e$
   k $\infty - \infty; 0$ **l** $\infty \times 0; 0$ **m** $\infty - \infty; 0$
   n $\infty \times 0; 1$ **o** $\infty^0; 1$
2. **a** $a = -2$ **b** $b = e^5$ **c** $c = \frac{1}{2}$
3. **a**

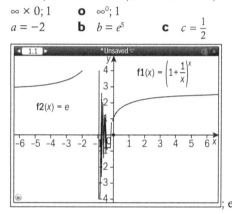

; e

   b $g(h) = \ln\left[(1+h)^{\frac{1}{h}}\right]$
   c $\lim_{h \to 0^+} g(h) = 1$; $\lim_{x \to +\infty} f(x) = e$

Exercise 2F

1. $a = 4, b = 3$ 2. $a = 3, b = \frac{9}{2}$ 3. No solutions

Exercise 2G

1. 0 2. $\frac{1}{2}$ 3. $\frac{1}{2}$

Review exercise

2. **b** e^{-2} 5. **a** $\frac{2}{\pi}$ **b** $\frac{4}{\pi}$
6. **d**

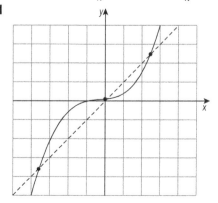

Chapter 3
Skills check

1. **a** $y = \ln\left(\frac{1-x^2}{x}\right)$ **b** $y = (x+1)e^{2x}$
   c $y = \frac{4x}{\pi}$ (assuming principal restriction)
   d $y = \frac{1 + \cos(2x)}{x}$
2. **a i** $(x^2 - 2x + 2)e^x + c$ **ii** $\frac{\sin(2x) - 2\cos(2x)}{5}e^x + c$
   iii $\ln|e^x - 1| + c$ **iv** $-\frac{1}{\sin x} + c$
   b i $y' = -\frac{1}{|x-1|\sqrt{1-2x}}$ **ii** $\frac{\sin(2x)}{\sin^2(x) + 1}$
3. **a** e^2 **b** $\frac{\sqrt{3}}{2}$

Exercise 3A

1.

Equation	Linear	Order	Coefficients
1	YES	3	Constant
2	NO	1	Constant
3	YES	2	Variable
4	NO	1	Constant
5	YES	1	Variable
6	YES	1	Variable
7	NO	1	Variable

2. **a** Yes **b** Yes **c** Yes

Investigation

1. $y = Ae^x$; $(0,3)$ contained in $y = 3e^x$
2. Expressions of the form $y = -\frac{1}{x+k}$; $(1, 1)$ in $y = -\frac{1}{x-2}$
3. No solution as $(y'(x))^2 \geq 0$

Exercise 3B

1. **a** $x = \frac{t^3}{3} + \frac{t^2}{2} + c$ **b** $y = \ln|x| + c|$
   c $z = x\sin(x) + \cos(x) + c$
   d $w = \frac{1}{2}x e^{2x} - \frac{1}{4}e^{2x} + c$
2. **a** $y = \frac{x^3}{3} + \frac{3x^2}{2} + \frac{13}{6}$ **b** $y = \sqrt{4 - x^2} - 1$
3. $y = \ln|x+2| + \frac{1}{2}\cos(x) + \frac{3}{2} - \ln 2$

5 $y = -\frac{1}{2}e^{-2x} - \ln|x-1| + \frac{3}{2}$

6 $s = \frac{1}{\sqrt{2}}\arctan\left(\frac{t}{\sqrt{2}}\right) + c$

Exercise 3C

1 a $\frac{1}{y-3}dy = dx;\ y = Ae^x + 3$

  b $(3y^2 + y - 1)\,dy = dx;\ y^3 + \frac{y^2}{2} - y = x + c$

  c $\frac{dy}{y^2 + 4} = dx;\ \frac{1}{2}\arctan\frac{y}{2} = x + c$

  d $\frac{1}{2y+1}dy = dx;\ \frac{1}{2}\ln|2y+1| = x + c$

  e $\frac{dy}{\cos^2(3y-1)} = dx;\ \frac{1}{3}\arctan(3y-1) = x + c$

  f $\frac{\cos y}{1+\sin^2 y}dy = dx;\ \arctan(\sin y) = x + c$

2 a $y = \sqrt{6e^x + c}$ **b** $y = Ax^{\frac{2}{5}} + 1$

  c $y = Ax$ **d** $y = -\tan(\arctan(x) + c)$

  e $y = \sqrt{4e^x(x-1) + c}$ **f** $y = \tan\left(-\frac{2}{x} + c\right)$

3 a $y = \sec^2 x$ **b** $y = \frac{x+2}{1-2x}$

  c $\ln|y-1| = 1 - e^{-x}$

4 a $\frac{dr}{dt} = \frac{2}{\pi r^2}$ **b** $r = \sqrt[3]{\frac{6}{\pi}(t-1) + 125}$

5 a No
  b and c

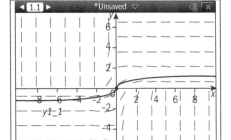

6 a

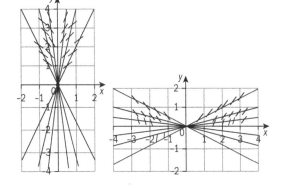

b

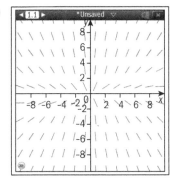

and

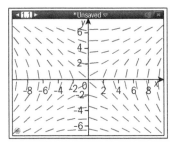

c

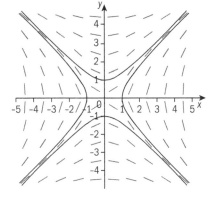

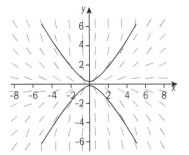

7 a

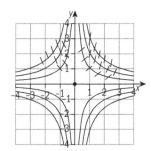

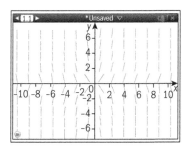

b

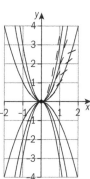

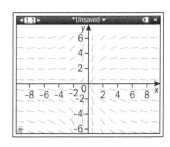

Short investigation

year	annually	monthly	daily	hourly	continously
40	4115.19	4211.61	4220.40	4220.68	4220.70

Exercise 3D

1 4800 years

2 4770 years

3 a $\dfrac{dV}{dt} = -0.00015V$

 b $V_0 = -0.00080024\ldots < 0.001$

4 $= \dfrac{1750 e^{\frac{t}{10}}}{43 + 7 e^{\frac{t}{10}}}$; approx. 50 years.

5 a 107.46 euros **b** 100×11^{I_0}
 c 116.18 euros

6 b $v = \dfrac{g}{\alpha} \dfrac{A e^{-\frac{1}{\alpha}t^2} - 1}{A e^{-\frac{1}{\alpha}t^2} + 1}$; $\lim_{t \to \infty} v(t) = \dfrac{g}{\alpha}$

Exercise 3E

1 a $y = \dfrac{1}{5} e^{2x} + c \cdot e^{-3x}$ **b** $y = e^{-x^2} + c \cdot e^{-x^2 - x}$

 c $y = \dfrac{1}{2} e^{-x^2}(\sin x - \cos x) + A e^{-x}$

 d $y = (x + c) \cos x$

 e $y = \left(-\dfrac{1}{4}\cos(2x) + c\right) \csc(x)$

2 a $y = x - 1 - e^{-x}$

 b $y = \dfrac{1}{5}\left(-2\sin(x) - \cos(x) + e^{2x}\right)$

 c $y = e^{-\frac{x^2}{2}} + e^{-\frac{x^2}{2} - x}$

3 All equations are both exact and linear.

 a $y = \dfrac{x^2}{3} + \dfrac{x}{2} + \dfrac{c}{x}$

 b $y = \dfrac{\sin x - \cos x}{2x + 2} e^x + \dfrac{c}{x + 1}$

 c $y = \left(\dfrac{3}{8}x + \dfrac{1}{4}\sin(2x) + \dfrac{1}{32}\sin(4x) + c\right)\sec(x)$

 d $y = \dfrac{\ln|\sec x| + c}{\sin x}$

4 a $y = A e^{-\frac{1}{x} + x}$ **b** $y = A\sqrt{x^2 + 1}$

 c $y = \dfrac{3}{2} + B e^{-x^2}$

5 $y = \dfrac{1}{1 - \dfrac{1}{2} e^{\frac{x^2}{2} - \frac{1}{2}}}$

Exercise 3F

1 Approx. 14°C

2 a $a = 5000$, $b = 0.04$
 b $x(0) = 1000$; $x = 1000 e^{0.04t} + 125000(e^{0.04t} - 1)$
 c 14.5 years. No

3 a $97\,ms^{-1}$
 b $491\,m$
 c $90\,ms^{-1}$

4 b 250 grams

Exercise 3G

1. **a** Homogeneous   **b** Non-homogeneous   **c** Non-homogeneous

2. **a** $y = \dfrac{x}{2}\left(3 + \dfrac{A}{x^2}\right)$

3. **a** $y = \dfrac{x}{3} + \dfrac{A}{\sqrt{x}}$ **b** $y = -x + Ax^3$

4. **a** $\dfrac{2}{\sqrt{15}} \arctan\left(\dfrac{2}{\sqrt{15}}\left(\dfrac{y}{x} - \dfrac{1}{2}\right)\right) = \ln|x| + c$

   b $Ay = e^{2y^2}$ **c** $\left(1 + \dfrac{y}{x}\right)^2 e^{\frac{y^2 - 2xy}{2x^2}} = Ax$

5. **a** $-\dfrac{1}{2}\left(\dfrac{x}{y}\right)^2 - \ln|y| + \ln 2 + 2 = 0$

   b $y = \dfrac{x \ln|x|}{\ln|x| + 1}$ **c** $\arctan\left(\dfrac{y}{x}\right) = \ln|x|$

6. **a**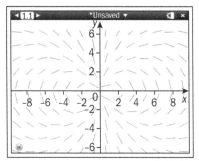

   b $\left(\dfrac{y}{x}\right)^2 + 1 = \dfrac{A}{\sqrt[3]{x^2}}$

   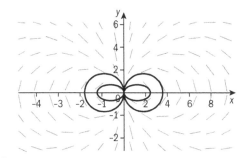

7. **f**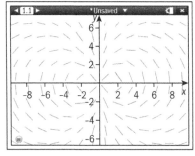

   $x^2 + y^2 + Ax + 1 = 0$; circles

8. **a** $\dfrac{dv}{du} = \dfrac{u - v}{u + v}$

   b $(x + 1)^2 - (y + 1)^2 - 2(x + 1)(y + 1) = c$

9. **a** $y = \dfrac{2}{1 + Ae^{2x}}$ **b** $y^2 = \dfrac{1}{Ae^{-2x} - 1}$

Investigation - approximation of Euler's Number

e.g approximations for h = 0.1, 0.01 and 0.001; the approximation of **e** gets better as the value of h decreases. However, for h = 0.0001, y(1) = 2.7105461… and this is a worse approximation of e.

	A n	B xn	C yn	D f
		=0+ 'n*0.1		
7	6	0.6	1.77156	1.77156
8	7	0.7	1.94872	1.94872
9	8	0.8	2.14359	2.14359
10	9	0.9	2.35795	2.35795
11	10	1.	2.59374	2.59374

C11 =c10+0.1·d10

	A n	B xn	C yn	D f
		=0+ 'n*0.01		
97	96	0.96	2.59927	2.59927
98	97	0.97	2.62527	2.62527
99	98	0.98	2.65152	2.65152
100	99	0.99	2.67803	2.67803
101	100	1.	2.70481	2.70481

C101 =c100+0.01·d100

	B xn	C yn	D f
	=0+'n*0.001		
997	996	0.996	2.70608	2.70608
998	997	0.997	2.70879	2.70879
999	998	0.998	2.7115	2.7115
100	999	0.999	2.71421	2.71421
1001	1000	1.	2.71692	2.71692

C1001 =c1000+0.001·d1000

Exercise 3H

1.

	A n	B xn	C yn	D f
		=0+'n*0.1		
7	6	0.6	1.77156	1.77156
8	7	0.7	1.94872	1.94872
9	8	0.8	2.14359	2.14359
10	9	0.9	2.35795	2.35795
11	10	1.	2.59374	2.59374

C11 =c10+0.1·d10

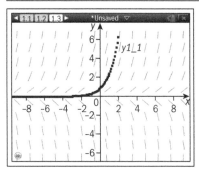

$y(1) = 2.59$

2 a
- $2x + y = 0$
- $2x + y = 1$
- $2x + y = -1$
- $2x + y = 2$
- $2x + y = -2$
- $2x + y = 3$
- $2x + y = -3>$

b and d

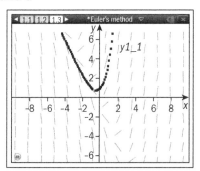

c

$y(1) = 4.1507 \ldots$ when $h = 0.001$

f $y = -2x - 2 + 2e^x; y(1) = 2e - 4$

g The approximation gives the correct answer to 3 s.f. (or 2 d.p.).

3 a

	A x	B y	C f	D
2	-1.8	1.6	0.68	
3	-1.6	1.736	-0.453696	
4	-1.4	1.64526...	-0.74688...	
5	-1.2	1.49588...	-0.79766...	
6	-1.	1.33635...	-0.78583...	
B6	=b5+0.2·c4			

$y(-1) = 1.34$

b

4 a

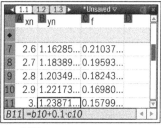

$y = 1.24$

b Exact value is $y = 1.23$

c e.g. Graphical method not reliable to predict solutions of the differential equation which are not close to initial value.

5 a

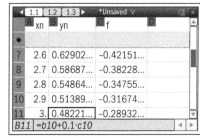

$y = 0.48$

b $f(x) = \dfrac{5}{1 + x^2}$

c 0.5. Approximation in part (a) is correct to 1 d.p.

6 a

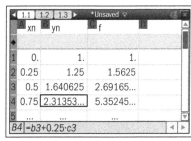

$y(0.25) = 1.25$
$y(0.5) = 1.64$
$y(0.75) = 2.31$
$y(1) = 3.64$

b $y(0.25) = 1.33$
$y(0.5) = 2$
$y(0.75) = 4$

c The error increased as the value of x increased.

7 a The temperature of a body decreases in proportion to the forth power of the body temperature.

b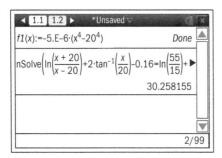

T(1) = 29.9

c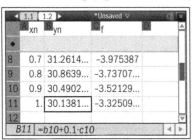

T(1) = 30.3

d Decrease the step (e.g. for h = 0.1, T(1) = 30.1)

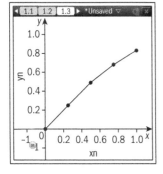

8 a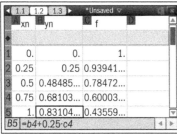

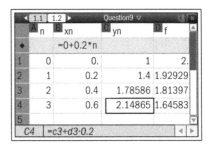

(0, 0), (0.25, 0.25), (0.5, 0.48), (0.75, 0.68), (1, 0.83).

b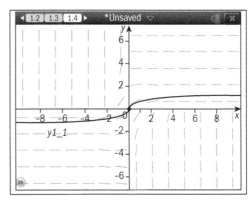

Review exercise

1 b i $A = 67$, $k = \frac{1}{10}\ln\frac{47}{67}$ **ii** 38.7

2 112

3 $y = \tan\left(\ln\left|\frac{x}{3}\right|\right)$

4 c $y = x - \dfrac{x}{\frac{1}{2}\ln|x| + C}$

5 $y = \frac{1}{2}\tan(x) + \left(\frac{1}{2}x + \frac{3}{4} - \frac{\pi}{8}\right)\sec^2(x)$

6 b $1 + \left(\frac{y}{x}\right)^2 = A|x|$ **c** $y = x\sqrt{5x-1}$, $x \geq \frac{1}{5}$

d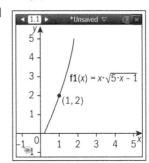

8 b $y = \tan x + 2\sec x$

9 a 2.149

b $y = \left(2\arcsin\frac{x}{2} + \frac{1}{2}\right)\sqrt{4-x^2}$

c 2.117

d

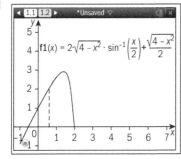

Explanation

10 a $V = V_0 e^{-\alpha t}$ **b** $t = \dfrac{\ln 3}{\alpha}$

11 $y = \pm \dfrac{\sqrt{x^2 - 4}}{2}$, $D =]-\infty, -2[\cup]2, \infty[$

12 $v = \pm \sqrt{\dfrac{k + mv_0^2}{m}}$

13 $a = -\dfrac{1}{4}$, $b = \dfrac{1}{4}$; $x = \dfrac{4Ae^{4kt}}{Ae^{4kt} - 1}$

14 b $y = \dfrac{1}{125l^3}\left(\dfrac{x^4}{12} - \dfrac{lx^3}{6} + \dfrac{l^3 x}{12}\right)$

    e $\dfrac{1}{1200}$; maximum distance below [AB]

Chapter 4

Skills check

1 a $L = 1$ **b** $L = \dfrac{3}{4}$;
   c $L = $ infinity   **d** $L = 1$

2 a Converges; $L = 1$
   b Converges $L = 0$
   c Converges $L = 0$

3 a $\dfrac{2}{3}\left((x+2)^{\frac{3}{2}} - x^{\frac{3}{2}}\right) + c$
   b $\arctan e^x + c$
   c $\sin(x) - x\cos(x) + c$

Exercise 4A

1 a Diverges by n^{th} term test
   b Converges; $S = 1.5$
   c Diverges by n^{th} term test
   d Converges; $S = 24.5$
   e Diverges by n^{th} term test
   f Diverges by n^{th} term test
   g Converges; $S_\infty = 6$
   h Converges; $S_\infty = 0.5$
   i Diverges by n^{th} term test

2 Converges to $\dfrac{3}{4}$

3 $x < -3$ or $x > 3$; $S_\infty = \dfrac{3x}{x - 3}$

Exercise 4B

3 a 25   **b** 4
4 a $\sqrt{3 + 4x}$ **b** $-2\tan(x)$ **c** $-x\sec(x)$
5 a 0   **b** 5   **c** $\dfrac{4}{5}$
6 $g'(x) = \sin(x)$
7 $x < -\dfrac{1}{2}$

Exercise 4C

1 Converges to 1   **2** Diverges
3 Converges to $\dfrac{\pi}{4}$ **4** Diverges
5 Diverges   **6** Converges to 2

Exercise 4D

1 Diverges   **2** Diverges
3 Converges   **4** Diverges
5 Converges
6 Integral test cannot be applied.
7 Converges
8 Integral test cannot be applied.
9 Integral test cannot be applied.
10 Diverges   **11** Converges
12 Converges   **13** Converges

Exercise 4E

1 Converges   **2** Diverges
3 Converges   **4** Diverges
5 Diverges

Exercise 4F

1 Converges by comparison with $\sum \dfrac{1}{n^3}$
2 Diverges by comparison with $\sum \dfrac{1}{\sqrt{n}}$
3 Converges by comparison with $\sum \dfrac{1}{n^2}$
4 Diverges by comparison with $\sum \dfrac{1}{n}$
5 Converges by comparison with $\sum \dfrac{1}{n^{\frac{3}{2}}}$
6 Diverges by comparison with $\sum \dfrac{1}{n}$
7 Converges by comparison with $\sum \dfrac{1}{n^2}$
8 Converges by comparison with $\sum \dfrac{1}{n^{\frac{3}{2}}}$
9 Converges by comparison to $\sum \dfrac{1}{n^{\frac{3}{2}}}$
10 Converges by comparison to $\sum \dfrac{1}{n^2}$

Exercise 4G

1. Diverges by comparison with $\sum \dfrac{1}{\sqrt{n^2}}$
2. Converges by comparison with $\sum \dfrac{1}{n^3}$
3. Diverges by comparison with $\sum \dfrac{1}{n}$
4. Converges by comparison with $\sum \dfrac{1}{2^n}$
5. Diverges by comparison with $\sum \dfrac{1}{n^{1/2}}$
6. Converges by comparison with $\sum \dfrac{1}{n^2}$
7. Diverges by comparison with $\sum \dfrac{1}{\sqrt{n}}$
8. Diverges by comparison with $\sum \dfrac{2^n}{n^2}$, which diverges by the nth term test

Exercise 4H

| 1 Diverges | 2 Converges | 3 Converges |
| 4 Converges | 5 Diverges | 6 Converges |

Exercise 4I

1. Converges absolutely
2. Converges absolutely
3. Converges absolutely
4. Converges absolutely
5. Converges absolutely
6. Does not converge absolutely

Exercise 4J

1. **a** Converges conditionally
   b Converges absolutely
   c Converges conditionally
   d Converges absolutely
   e Converges absolutely
   f Converges conditionally
   g Converges conditionally
   h Converges absolutely
2. 0.902   3. 0.948
4. $u_s = 0.000033$; $S = 0.012644$
5. $n = 29$

Exercise 4K

1 Converges	2 Converges	3 Converges
4 Converges	5 Diverges	
6 Converges conditionally		
7 Diverges	8 Diverges	9 Diverges
10 Converges	11 Converges	12 Diverges
13 Converges	14 Converges	

Review exercise

1. **a** Diverges **b** Diverges **c** Diverges
   d Converges **e** Converges
2. $S = 1$
3. 0.992594
6. **a** Converges **b** Converges
7. Converges
9. **c i** 1.082   **ii** $N = 90$
12. 0.841468; $\dfrac{1}{9!}$

Chapter 5

Skills Check

1. **a** $-2 < x < 2$
   b $-\dfrac{2}{3} < x < \dfrac{2}{3}$; $1 + \dfrac{3}{2}x + \dfrac{9}{4}x^2 + \ldots + \left(\dfrac{3}{2}x\right)^{n-1} + \ldots$
2. **a** $\dfrac{2}{1 + 4x^2}$ **b** $\dfrac{3}{3x-1}$ **c** $e^{\sin x} \cos x$
3. **a** $\sin x - 2e^{-\frac{x}{2}} + C$ **b** $\dfrac{3}{2}\arctan\left(\dfrac{x}{2}\right)$
   c $-\dfrac{1}{4}\ln(1 - 4x)$

Exercise 5A

1. **a** $\sum\limits_{x=0}^{\infty}(4x)^n$; $\left]-\dfrac{1}{4}, \dfrac{1}{4}\right[$; $R = \dfrac{1}{4}$
   b $\sum\limits_{x=0}^{\infty}(-1)^n(5x)^n$; $\left]-\dfrac{1}{5}, \dfrac{1}{5}\right[$; $R = \dfrac{1}{5}$
   c $-\dfrac{1}{4}\sum\limits_{x=0}^{\infty}\left(\dfrac{1}{4}x\right)^x$; $]-4, 4[$; $R = 4$
   d $\sum\limits_{n=0}^{\infty}(-1)^n(3x-1)^n$; $\left]0, \dfrac{2}{3}\right[$; $R = \dfrac{1}{3}$
   e $x\sum\limits_{n=0}^{\infty}(-1)^n x^n = \sum\limits_{n=0}^{\infty}(-1)^n x^{n+1}$; $]-1, 1[$; $R = 1$
   f $\sum\limits_{x=0}^{\infty}2(-1)^n(4x)^n$; $\left]-\dfrac{1}{4}, \dfrac{1}{4}\right[$; $R = \dfrac{1}{4}$

Exercise 5B

1. **a** $R = 1$; $]-1, 1]$ **b** All reals
   c $R = \dfrac{1}{2}$; $\left[-\dfrac{1}{2}, \dfrac{1}{2}\right]$ **d** $R = \dfrac{1}{2}$; $]-2, -1]$
   e $R = 1$; $]0, 2]$ **f** All reals
   g $R = 3$; $[-3, 3]$ **h** $R = 3$; $]-3, 3[$
   i All real x
   j $x < 0$ **k** $R = 3$; $]-4, 2[$
2. **a** $k = 2$ **b** $k = 3$
3. $h = 4, k = 3$

Exercise 5C

1 $2\sum_{n=0}^{\infty} x^{2n}$; $]-1, 1[$; $R = 1$

2 $\sum_{n=0}^{\infty}(-1)^n nx^{n-1}$; $]-1, 1[$; $R = 1$

3 $\sum_{n=0}^{\infty}(-1)^n 4^n x^{2n}$; $\left]-\frac{1}{2}, \frac{1}{2}\right[$; $R = \frac{1}{2}$

4 $1 + x + \frac{x^2}{2} + \frac{3x^4}{8} + \ldots$; $]-1, 1[$; $R = 1$

5 $\sum_{n=1}^{\infty}(-1)^n \frac{(x-1)^{n+1}}{n+1} + C$; $I = [0, 2]$; $R = 1$

6 1

7 $[-2, 2]$ and $R = 2$; $[-2, 2[$ and $R = 2$;
One is the derivative of the other, so the interval is the same, apart from the endpoints.

8 $\sum_{n=0}^{\infty}\left[(-1)^n\left(\frac{x}{2}\right)^n - x^n\right]$; $]-1, 1[$; $R = 1$

Exercise 5D

1 a $T_4(x) = 1 + x + x^2 + x^3 + x^4$

  b $T_4(x) = x^2 - \frac{x^4}{6}$

  c $T_4(x) = 1 + \frac{x}{3} - \frac{x^2}{9} + \frac{5x^3}{81} - \frac{10x^4}{243}$

  d $T_4(x) = x + x^2 + \frac{x^3}{2} + \frac{x^4}{6}$

2 a $T_3(x) = \frac{1}{\pi} - \frac{1}{\pi^2}(x - \pi) + \frac{1}{\pi^3}(x - \pi)^2 - \frac{1}{\pi^4}(x - \pi)^3$

  b $T_3(x) = -\frac{1}{\pi}(x - \pi) + \frac{1}{\pi^2}(x - \pi)^2 + \frac{\pi^2 - 6}{6\pi^3}(x - \pi)^3$

  c $T_3(x) = -\pi^2 - 2\pi(x - \pi) + \left(\frac{\pi^2 - 2}{2}\right)(x - \pi)^2 + \pi(x - \pi)^3$

3 a $T_3(x) = -5 + 4x - 3x^2 + 2x^3$

  b $T_3(x) = -2 + 4(x - 1) + 3(x - 1)^2 + 2(x - 1)^3$

4 The Taylor Polynomial of degree 1 matches the function and its first derivative at $x = a$.

5 $T_3(x) = 4 - (x - 1) + \frac{3}{2}(x - 2)^2 + \frac{1}{3}(x - 1)^3$
$f(1.2) = 3.863$

Exercise 5E

2 a $\sin^2 x = \frac{2x^2}{2!} - \frac{8x^4}{4!} + \frac{32x^6}{6!} + \ldots + \frac{(-1)^{n+1}2^{2n-1}x^{2n}}{(2n)!} + \ldots$

  b $e^{3x} = 1 + 3x + \frac{9x^2}{2!} + \ldots + \frac{3^n x^n}{n!}$

  c $\ln(1 - x) = -x - \frac{x^2}{2} - \frac{x^3}{3} - \frac{x^4}{4} \ldots - \frac{x^n}{n} - \ldots$

  d $\tan x = x + \frac{1}{3}x^3 + \frac{2}{15}x^5 + \frac{17}{315}x^7 + \ldots$

  e $\arctan x^2 = x^2 - \frac{x^6}{3} + \frac{x^{10}}{5} - \frac{x^{14}}{7} + \ldots + (-1)^n \frac{x^{2(2n+1)}}{2n+1} + \ldots$

  f $e^x \sin x = x + x^2 + \ldots$

3 $\sin^2 x = \frac{1}{2} - \frac{1}{2}\cos x = \frac{2}{2!}x^2 - \frac{2^3}{4!}x^4 + \frac{2^5}{6!}x^6 - \frac{2^7}{8!}x^8 + \ldots$
(Series is the same as the one found in **2a**.)

4 $\sum_{n=0}^{\infty}(-1)^n \frac{x^{2n}}{(2n+1)!}$

Exercise 5F

1 a $1 - \sum_{n=1}^{\infty} \frac{(2n)!}{(2n \cdot n!)^2(2n - 1)} x^n$; $|x| \leq 1$

  b $\sum_{n=0}^{\infty}(-1)^n \frac{(n+1)(n+2)}{2} x^n$; $|x| < \frac{1}{2}$

  c $\sum_{n=0}^{\infty} 4^n(n+1)x^{2n}$; $|x| < 1$

  d $1 + \frac{x^3}{2} + \sum_{n=2}^{\infty}(-1)^{n-1} \frac{3 \cdot 7 \cdots (4n-5)}{2^n \cdot n!} x^{3n}$; $|x| < \frac{1}{\sqrt[3]{2}}$

2 $\arcsin x = x + \frac{1}{2 \cdot 3}x^3 + \frac{1 \cdot 3}{2 \cdot 4 \cdot 5}x^5 + \frac{1 \cdot 3 \cdot 5}{2 \cdot 4 \cdot 6 \cdot 7}x^7 + \ldots$
$= \sum_{n=0}^{\infty} \frac{(2n)!}{2^{2n}(n!)^2(2n+1)} x^{2n+1}$

3 a 3.017 **b** 0.8885

Exercise 5G

1 $e^{-1} \approx 0.368$

2 0.4055

3 2.7183

4 6.2337×10^{-7}

5 Degree 4, $T_5\left(\frac{1}{3}\right) = 1.3956$

6 $T_2(x) = 2 + \frac{1}{12}(x - 8) - \frac{1}{288}(x - 8)^2$;
accurate within 0.0003

7 0.045293; Error of 10^{-6}

8 0.0000217

9 3×10^{-6}

10 5

Exercise 5H

1. a) $-\dfrac{1}{6}$ b) 2   c) 1
   d) $\dfrac{1}{3}$ e) $-\dfrac{162}{15625}$

2. $y = 1 + x + \dfrac{1}{2}x^2 + \dfrac{4}{3!}x^3 + \dfrac{14}{4!}x^4 + \dfrac{66}{5!}x^5 + \ldots$;
   $y(0.2) = 1.2264$

3. $y = 1 + 3(x-1) + 7.5(x-1)^2 + 14.5(x-1)^3$

4. $y = 1 + x + \sum_{n=2}^{\infty}(n-1)!\,x^n$; no solution since this series only converges for $x = 0$.

Review exercise

1. $r = \dfrac{1}{4}$

2. $1 - x - \dfrac{3}{2}x^2 + \dfrac{11}{6}x^3 - \dfrac{7}{24}x^4$

3. a Converges
   b $\sin x = x - \dfrac{x^3}{3!} + \dfrac{x^5}{5!} - \dfrac{x^7}{7!}$; $e^{x^2} = 1 + x^2 + \dfrac{x^4}{2!} + \dfrac{x^6}{3!}$
   c $x + \dfrac{5}{6}x^3 + \dfrac{41}{120}x^5$ d $\dfrac{5}{6}$

4. a ii $a_n = \dfrac{1 \cdot 3^2 \cdots (n-2)^2}{n!}$
   b $R = 1$ c 3.139

5. $[1, 5]$
6. 1.221

7. $x + \dfrac{x^3}{3} + \dfrac{x^5}{5} + \ldots$

8. 1.06272

9. $\sin(3°) = \sin\left(\dfrac{\pi}{60}\right) \approx \dfrac{\pi}{60} - \dfrac{\left(\dfrac{\pi}{60}\right)^3}{3!} \approx 0.05234$

10. a $f^{(n)}(x) = \dfrac{e^x + (-1)^n e^{-x}}{2}$
    b $f(x) = 1 + \dfrac{x^2}{2} + \dfrac{x^4}{24} + \ldots$
    c $f\left(\dfrac{1}{2}\right) = \dfrac{433}{384}$; 0.000136

11. b $x - \dfrac{x^2}{2} + \dfrac{x^3}{6} - \dfrac{x^4}{12} + \ldots$
    c $-x - \dfrac{x^2}{2} - \dfrac{x^3}{6} - \dfrac{x^4}{12} + \ldots$
    e $L = 0$

13. a $y = -\dfrac{\pi}{2} + x - \dfrac{\pi}{4}x^2$
    b $y = \left(\dfrac{x}{2} + \dfrac{\sin 2x}{4} - \dfrac{\pi}{2}\right).\sec x$

14. $1 + \dfrac{3}{2}x + \dfrac{15}{8}x^2 + \dfrac{51}{16}x^3 + \ldots$; $\left(-\dfrac{1}{2}, \dfrac{1}{2}\right)$

15. $1 + x + x^2 + \dfrac{5}{6}x^3$; $\dfrac{1}{2}$

16. $n = 40{,}001$

17. 0.341354

Index

Page numbers in *italics* indicate answers sections.

A

Abel, Niels Henrik 131
absolute convergence of series 120–2
d'Alembert, Jean le Rond 119
algebra of limits 7–9, 21
algorithms 86–7, 95
alternative series test 122–5
Antiphon of Athens 104
Archimedes 104

B

Basel Problem 101
Bernoulli Equations 84
Bernoulli, Jakob 101
Bernoulli, Johann 57, 73
Binomial Series 152–5
Blancmange function 23
Bolzano's Theorem 53
boundary conditions 60
Bryson of Heraclea 104
Butterfly Effect 89

C

calculus 55, 97
  fundamental theorem of calculus (FTC) 107–9, 129
Cantor, Georg 3
carbon dating 70–1
Cauchy Criterion 97
Cauchy, Augustin-Louis 13, 24, 108
Cauchy's Theorem 42–3, 53
Cesàro series 98
Cesàro sum 98
Cesàro, Ernesto 98
Chain Rule 52
Chaos Theory 89
comparison test for convergence 115–18
conditional convergence of series 122–5
continuous compound interest 69–70, 78–9
continuous functions 23
  composition of continuous functions 28
  continuity and differentiability on an interval 24–8
  theorems about continuous functions 28–32, 52
continuum hypothesis 3
contrapositive statement 100

convergence 96, *181–2*
  absolute convergence of series 120–2
  comparison test for convergence 115–18
  conditional convergence of series 122–5
  convergence of an improper integral 110–12
  convergence of infinite series 97
  convergent series 98
  integral test for convergence 112–14, 129
  introduction to convergence tests for series 104–9
  limit comparison test for convergence 118–19
  p-series test 114–15
  ratio test for convergence 119–20
  review exercise 127–8
  series and convergence 98–103
  summary of tests for convergence 126–7
convergent sequences 5, 21
  subsequences of convergent and divergent sequences 5–7, 21
Cramer's Rule 148
criteria for maxima/minima 53

D

d'Alembert ratio 119
definite integral 106, 129
derivative of inverse function 53
differentiable functions 23, 33–7
  continuity and differentiability on an interval 24–8
  Mean Value Theorem 38–42
  Rolle's theorem 37–8
  theorems about differentiable functions 52–3
differential equations 54, 55–6, 95, *175–81*
  classifications of differential equations and their solutions 56–8
  differential equations and Taylor Series 162–3
  differential equations with separated variables 61–3
  Euler Method for first order differential equations 85–91
  first order exact equations and integrating factors 73–6

homogeneous differential equations and substitution methods 80–4
  initial values or boundary conditions 60
  modeling of growth and decay phenomena 69–72
  order of equation 57
  ordinary differential equations (ODE) 57
  real life applications of first order linear differential equations 76–80
  review exercise 92–4
  separable variables differential equations and graphs of their solutions 63–9
  solutions 58–61
divergence 98, 129
  nth term test for divergence 100, 129
divergent sequences 10–13
  subsequences of convergent and divergent sequences 5–7, 21

E

Euler, Leonhard 73, 97, 98, 101
Euler's Method 87–91, 95
  approximation of Euler's number 89
exact equations 73–4
exponential decay curve 70
exponential growth curve 69
Extreme Value Theorem 107

F

falling body and Newton's second law 77
first order exact equations 73–6
first order linear differential equations 76–80, 95
  Euler Method for first order differential equations 85–91
fundamental theorem of calculus (FTC) 107–9, 129

G

Gabriel's horn 111
Gauss, Carl Friedrich 17, 24
graphs 64–5, 95
  isoclines and graphical approximations of solutions 65–9
  smooth graphs of functions 49–50

Gregory, James 148
growth and decay phenomena 69–72
   exponential decay curve 70
   exponential growth curve 69

H

harmonic series 101
Heine, Heinrich Eduard 17
Heine's definition 13
Hilbert, David 97
homogeneous differential equations 80–4, 95

I

improper integrals 110–12
   improper integral notation 110, 129
indeterminate forms 46–8, 53
infinite sequences 2, *168–9*
   convergent sequences 5, 21
   definition of right limit at a point 15
   divergent sequences 10–13
   from limits of sequences to limits of functions 3–4, 13–19
   review exercise 20
   squeeze theorem and the algebra of limits of convergent sequences 7–9, 21
   subsequences of convergent and divergent sequences 5–7, 21
infinity 3
initial values 60
integral as the limit of sums 104–7
   improper integrals 110–12
integral test for convergence 112–14, 129
integrating factors 73–6
interval of convergence 132, 133, 166
isoclines 65–9, 95

J

Jyesthadeva 145

K

Kerala School, India 97

L

L'Hôpital's Rule 43–5, 53
   power series 161
LaGrange's form 149, 167
Laplace, Pierre Simon 97
Leibniz 57, 60, 97, 101
Leibniz's Theorem 122–5
Libby, Willard Frank 70
limit comparison test for convergence 118–19
limits of sequences to limits of functions 13–19, 50

definition of left limit at a point 15
definition of right limit at a point 15
Lorenz Attractor 89
lower bound sums 105

M

Maclaurin Series 146–55, 166
Maclaurin, Colin 148
Madhava of Sangamagrama 145
Mean Value Theorem 38–42, 53
Mengoli, Pietro 101
Mercator, Nicolaus 148
method of exhaustion 104
Method of Tangents 87, 95
mixing solutions problems 78

N

Newton, Sir Isaac 57, 97
Newton's cooling law 76–7
Newton's second law 77

O

order of equation 57, 95
Oresme 101

P

p-series test 114–15
partial sums 98–103
pathological functions 23, 35
Poincaré, Henri 23
polynomials 130, *182–4*
   differential equations and Taylor Series 162–3
   from finite to 'infinite' polynomials 131
   representing functions by power series 1 132–5
   representing functions by power series 2 138–43
   representing power series as functions 135–8
   review exercise 164–5
   Taylor and Maclaurin Series 146–55
   Taylor polynomials 143–6
   useful applications of power series 161
   using Taylor Series to approximate functions 156–60
power series 1 132–5
power series 2 138–43
power series and L'Hôpital's Rule 161
power series as functions 135–8
power series centered at $x = 0$ 132, 166

R

radius of convergence 132, 133, 166
ratio test applied to power series 136–7, 166–7
ratio test for convergence 119–20
Riemann sums 105–7, 129
Rolle's Theorem 37–8, 53
Russell, Bertrand 3

S

Sandwich Theorem 7
slope fields 64, 95
smooth graphs of functions 49–50
smoothness in mathematics 22, *169–75*
   continuity and differentiability on an interval 24–8
   differentiable functions 33–42, 52–3
   exploring continuous and differentiable functions 23
   limits at a point, indeterminate forms and L'Hôpital's Rule 42–48, 53
   limits of function and limits of sequences 50
   review exercise 51
   theorems about continuous functions 28–32, 52
   what are smooth graphs of functions? 49–50
squeeze theorem 7–9, 21
subsequences of convergent and divergent sequences 5–7, 21
substitution methods 80–4

T

Taylor polynomials 143–6
Taylor Series 146–55
   differential equations and Taylor Series 162–3
   using Taylor Series to approximate functions 156–60
Taylor, Brook 97, 148
Taylor's Formula 149, 167
Torricelli, Evangelista 111
Torricelli's trumpet 111

U

upper bound sums 105–6

V

Vanishing Condition 100

W

Weierstrass, Karl 131
Weierstrass' Theorem 30, 35, 53, 107